Guide to Multiple Regression

This book is intended to equip the reader with concept knowledge on multi-linear regression methods, and develop technical skills while solving problems with the R programming language. The goal is not to become an expert in both directions. It is rather an incentive to do research on advanced models in data analysis once confidence is gained with the practice of some methods.

Basics in inferential statistics and basics in a programming language are necessary prerequisites to engage well with the content. We believe that using a computer is the only practical means of performing a linear or multilinear regression on a data set of even moderate size. Therefore, our purpose is to make these connections between theory, technology, and applications as clear as possible.

Features

- Focused summaries of the main statistical methods, followed by solved questions
- Integration of R as a calculator and as a programming language in solving each question with ease and accuracy
- Use of simple coding that illustrates the connection to the theory
- Suggest progressively alternative codes
- Visualization and interpretation of the R outputs

Samia Challal is an associate professor at Glendon College, the bilingual campus of York University. She teaches courses in Analysis, Algebra, and Statistics. Her research draws on partial differential equations applied to fluid dynamics, optimization, homogenization problems, and mathematics education. She received her PhD from the University of Metz, France.

Guide to Multiple Regression

Samia Challal

CRC Press
Taylor & Francis Group
Boca Raton London New York

CRC Press is an imprint of the
Taylor & Francis Group, an **informa** business

A CHAPMAN & HALL BOOK

Designed Cover Image: Samia Challal

First edition published 2026
by CRC Press
2385 NW Executive Center Drive, Suite 320, Boca Raton FL 33431

and by CRC Press
4 Park Square, Milton Park, Abingdon, Oxon, OX14 4RN

CRC Press is an imprint of Taylor & Francis Group, LLC

ISBN: 978-1-041-01891-9 (hbk)
ISBN: 978-1-041-01892-6 (pbk)
ISBN: 978-1-003-61683-2 (ebk)

DOI: 10.1201/9781003616832

Typeset in Latin Modern font
by KnowledgeWorks Global Ltd.

Publisher's Note: This book has been prepared from camera-ready copy provided by the author.

To my parents

Contents

List of Figures

List of Tables

Preface

Goal

The focus of this content is on multi-linear regression methods. Its purpose is to acquire some knowledge of the basic methods and some technical skills by learning the implementation of these methods using the R programming Language. The goal is not to become an expert in both directions. It is rather an incentive to do research on advanced models in data analysis once confidence is gained while manipulating basic methods. The reader can explore Logistic Regression, Multinomial Regression, Count Regression in (S. Chatterjee, 2013), a variety of Multilevel Regression in (A. Gelman, 2006), and Robust Regression, Fuzzy Regression, and more Regression methods in (Panik, 2010) and (P. Roback, 2021).

Prerequisites

We assume that the tools on basic probabilities, distributions, estimations, and hypothesis testing are well known and that the reader is comfortable with different notations and terminologies (Gut, 1995; R. Bartoszynski, 2008; William Mendenhall, 2018; R.J. Larsen, 2001; Panik, 2012; Larson, 1982).

We also, assume that the reader has at least some knowledge with a programming language. A quick initiation of R can be done by installing R (r project.org, 2 24) and RStudio IDE (integrated development environment) (posit.co/download/rstudio desktop, 2 24) and then getting familiar with R objects (STHADA, 2 24). For more detailed R content, we refer the reader to the references (Lander, 2014; Chang, 2013; Crawley, 2013; Gardener, 2012).

R is available as Free Software. It compiles and runs on a wide variety of platforms. It provides a wide variety of statistical and graphical techniques. "R has its own LaTeX-like documentation format, which is used to supply comprehensive documentation, both online in several formats and in hardcopy" (r project.org, 2 24). In particular, working on R Markdown (rmarkdown.rstudio.com, 2 24) (Yihui Xie, 2023) files offers this possibility.

Structure of the Book and Pedagogical Approach

In this book, we guide the reader, step by step through:

- the theoretical concepts without focusing on the proofs,

- the illustration of the concepts by concrete examples,

- the implementation of the developed methods using the R-programming language, where the codes are introduced progressively.

In each section, a short summary of the theoretical tools is given, followed by references for further and deep exploration. Then, a set of solved examples serves as models for the reader. R coding is integrated and shown in the solution. This initiation to R lessens the complexity of calculations and facilitates visualization. However, dealing with a new technology adds its own challenges. That is why we prioritize the use of simple coding that illustrates the connection to the theory. Once we get familiarized with the coding used, we suggest progressively alternative codes that do the same work by calling specific R-functions.

We hope that this approach will support the learning curve. On one side, by realizing the power of R tools, the learner gains confidence and then is encouraged to do their own research for R codes. On the other hand, this approach will equip the learner with two logical strategies while solving problems with R;

- understanding the theory is a necessary step to implement a statistical analysis with any software,

• understanding what does a particular code is necessary to do the targeted calculation.

Our motivation in preparing this guide is because we believe that using a computer is the only practical means of performing a linear or multi-linear regression on a data set of even moderate size. Therefore, our purpose is to make these connections between theory, technology, and applications as clear and simple as possible.

Software information and conventions

I used the **knitr** package (Xie, 2015) and the **bookdown** package (Xie, 2024) to compile my book. My R session information is shown below:

```
xfun::session_info()
```

```
## R version 3.6.2 (2019-12-12)
## Platform: x86_64-w64-mingw32/x64 (64-bit)
## Running under: Windows 10 x64 (build 19045)
##
## Locale:
##   LC_COLLATE=English_Canada.1252
##   LC_CTYPE=English_Canada.1252
##   LC_MONETARY=English_Canada.1252
##   LC_NUMERIC=C
##   LC_TIME=English_Canada.1252
##
## Package version:
##   base64enc_0.1.3    bookdown_0.43
##   bslib_0.2.4        compiler_3.6.2
##   digest_0.6.27      evaluate_1.0.3
##   fontawesome_0.5.3  fs_1.5.0
##   graphics_3.6.2     grDevices_3.6.2
##   highr_0.11         htmltools_0.5.1.1
##   jquerylib_0.1.4    jsonlite_1.7.2
##   knitr_1.50         magrittr_2.0.1
##   methods_3.6.2      R6_2.6.1
```

```
##   rappdirs_0.3.3    rlang_0.4.12
##   rmarkdown_2.29    rstudioapi_0.17.1
##   sass_0.4.9        stats_3.6.2
##   tinytex_0.57      tools_3.6.2
##   utils_3.6.2       xfun_0.52
##   yaml_2.2.0
```

Package names are in bold text (e.g., **rmarkdown**), and inline code and filenames are formatted in a typewriter font (e.g., `knitr::knit('foo.Rmd')`). Function names are followed by parentheses (e.g., `bookdown::render_book()`).

Acknowledgments

I have relied on the various authors cited in the bibliography, and I am grateful to all of them. Many exercises are drawn or adapted from the cited references for their aptitude to reinforce the understanding of the material.

Samia Challal

1

Simple linear probabilistic model

1.1 Description

Suppose we pick up n values: $x_1, \ldots, x_n$ (MT: Motivational Test scores) and for each x_i, we select one or more at random responses $y_{x_i} = y_i$ (CS: Calculus average Scores). The simplest possible way in which we could model a variation of the response with respect to the predictor variable x is to assume, in particular, that the mean of the random variable Y_x is a function of x; that is, $E(Y_x) = g(x)$, where g is a suitable function. In order to identify g, the Least Squares Method (LSM) is adopted. The LSM minimizes the deviation between the observed values y_i and the means $g(x_i)$:

$$e_i = Y_i - g(x_i)$$

called an error of observation. More precisely, LMS is described as (Larson, 1982):

Definition. Let $Y_1, Y_2, \ldots, Y_n$ be uncorrelated random variables,

$$
\begin{Vmatrix}
E(Y_i) = g(x_i) & Y_i = Y_{x_i} & Var(Y_i) = \sigma^2 & i = 1, \ldots, n \\
x_1, \ldots, x_n & \text{are constants} & \\
g & \text{is a function involving unknown parameters}
\end{Vmatrix}
$$

where $E(Y_i)$ is the expected mean of Y_i and $Var(Y_i)$ is the variance of Y_i.

The least squares estimators of the unknown parameters in $g(x)$ are those values that minimize

$$Q = \sum_{i=1}^{n} (Y_i - g(x_i))^2 = \sum_{i=1}^{n} e_i^2$$

DOI: 10.1201/9781003616832-1

The estimator for σ^2 is

$$S^2 = k \sum_{i=1}^{n} (Y_i - \widehat{g(x_i)})^2$$

where $\widehat{g(x_i)})$ is the least squares estimator for $E(Y_i) = g(x_i)$ and k is chosen to make S^2 unbiased $(E(S^2) = 0)$.

In particular, for a linear model, we establish the following (Larson, 1982):

Theorem. Let $Y_1, Y_2, \ldots, Y_n$ be uncorrelated random variables,

$$\left\|\begin{array}{l} E(Y_i) = a + bx_i \\ x_1, \ldots, x_n \end{array}\right. \qquad \begin{array}{l} Var(Y_i) = \sigma^2 \qquad i = 1, \ldots, n \\ \text{are constants that are not all equal.} \end{array}$$

The least squares estimators for a and b are $\widehat{A}$ and $\widehat{B}$ given by

$$\widehat{A} = \overline{Y} - \widehat{B}\overline{x} \qquad\qquad \widehat{B} = \frac{\displaystyle\sum_{i=1}^{n} (x_i - \overline{x})(Y_i - \overline{Y})}{\displaystyle\sum_{i=1}^{n} (x_i - \overline{x})^2}$$

where

$$\overline{x} = \frac{1}{n} \sum_{i=1}^{n} x_i \qquad\qquad \overline{Y} = \frac{1}{n} \sum_{i=1}^{n} Y_i.$$

The estimator for σ^2 is

$$S^2 = \frac{1}{n-2} \left[\sum_{i=1}^{n} (Y_i - \overline{Y})^2 - \widehat{B}^2 \sum_{i=1}^{n} (x_i - \overline{x})^2 \right].$$

Idea of the proof. The point (a, b) that minimizes the function

$$Q(a, b) = \sum_{i=1}^{n} (y_i - a - bx_i)^2$$

is the solution of the system

$$\frac{\partial Q}{\partial a} = -2\sum_{i=1}^{n}(y_i - a - bx_i) = 0$$

$$\frac{\partial Q}{\partial b} = -2\sum_{i=1}^{n}x_i(y_i - a - bx_i) = 0.$$

We show that Q is strictly convex and that the unique critical point (a, b) is the global minimum (Challal, 2020).

Notation. We define the statistics

$$S_{yy} = \sum_{i=1}^{n}(Y_i - \overline{Y})^2 = \sum_{i=1}^{n}Y_i^2 - \frac{1}{n}\left(\sum_{i=1}^{n}Y_i\right)^2$$

$$S_{xy} = \sum_{i=1}^{n}(x_i - \overline{x})(Y_i - \overline{Y}) = \sum_{i=1}^{n}x_iY_i - \frac{1}{n}\left(\sum_{i=1}^{n}x_i\right)\left(\sum_{i=1}^{n}Y_i\right),$$

and the quantity

$$s_{xx} = \sum_{i=1}^{n}(x_i - \overline{x})^2 = \sum_{i=1}^{n}x_i^2 - \frac{1}{n}\left(\sum_{i=1}^{n}x_i\right)^2 \qquad \overline{x} = \frac{1}{n}\sum_{i=1}^{n}x_i.$$

If $y = (y_1, y_2, \ldots, y_n)$ is an observed sample corresponding to $x = (x_1, x_2, \ldots, x_n)$, then the values of the random variables S_{yy}, S_{xy} are

$$s_{yy} = \sum_{i=1}^{n}(y_i - \overline{y})^2 = \sum_{i=1}^{n}y_i^2 - \frac{1}{n}\left(\sum_{i=1}^{n}y_i\right)^2$$

$$s_{xy} = \sum_{i=1}^{n}(x_i - \overline{x})(y_i - \overline{y}).$$

The values of the point estimators $\widehat{A}$, $\widehat{B}$, and S^2 are respectively $\widehat{a}$, $\widehat{b}$, and s^2 given by

$$\widehat{a} = \overline{y} - \widehat{b}\overline{x} \qquad \widehat{b} = \frac{s_{xy}}{s_{xx}} \qquad s^2 = \frac{1}{n-2}\left[s_{yy}^2 - \frac{s_{xy}^2}{s_{xx}}\right].$$

Remark. We have

- $$E(e_i) = 0 \qquad \text{and} \qquad Var(e_i) = \sigma^2$$

- The Simple Linear Regression Model has three unknown parameters: $a, b,$ and σ^2.

- The y-intercept a represents the predicted value of y when $x = 0$.

- The slope b represents the increase or decrease in y for every 1-unit increase in x.

TABLE 1.1 MT and CS scores

MT	CS	MT	CS
92.76	89.24	64.90	76.45
72.64	65.79	60.12	64.58
86.72	76.15	66.62	64.13
94.08	90.44	70.76	71.71
63.56	70.77	86.34	81.05
83.42	70.15	68.20	71.76

Solved Problems

1.1.1 Effect of a Motivational Test on Calculus Scores

In order to improve first-year students' experience in a Calculus course, an instructor assigned modules for reviewing basics of mathematics that are prerequisites to the Calculus course for a period of 2 to 3 weeks. At the end of the review, students have to take a test called "MT". The purpose of the "Motivational Test" MT is to motivate students to take their review seriously and make a responsible decision on whether they continue the Calculus course or transfer to a Precalculus Course. The MT scores and Calculus scores "CS" are recorded in Table 1.1.

Find the least-squares prediction line for calculus grade data.

Solution. Let x and y be the vectors of the MT and CS scores, respectively. The regression line of best fit is given by $y = a + bx$ where a and b are defined as follows:

$$s_{xy} = \sum_{i=1}^{n} (x_i - \overline{x})(y_i - \overline{y}), \qquad s_{xx} = \sum_{i=1}^{n} (x_i - \overline{x})^2$$

$$b = \frac{s_{xy}}{s_{xx}}, \qquad a = \overline{y} - b\overline{x}.$$

```
x=c(92.76, 64.90,72.64 ,60.12 ,86.72, 66.62, 94.08, 70.76,
                          63.56, 86.34, 83.42, 68.20)
y=c(89.24, 76.45, 65.79, 64.58, 76.15, 64.13, 90.44, 71.71,
                          70.77, 81.05, 70.15, 71.76)


sxy=sum((x-mean(x))*(y-mean(y)))
sxx=sum((x-mean(x))*(x-mean(x)))


b = sxy/sxx                           # slope
b
```

```
## [1] 0.5823
```

```
a = mean(y)- b*mean(x)                 # intercept
a
```

```
## [1] 30.19
```

Comments. The equation of the line is: $y = a + bx = 30.18745 + 0.5823085x$.

The intercept $a = 30.18745$ is the final mean average that would have a student who doesn't do, for example, the placement test and gets 0 (see Figure 1.1).

The slope $b = 0.5823085 = (y - y_0)/(x - x_0)$ is the amount of increase in the final average for every one unit increase in the MT grade.

```
plot(x,y,  col = "blue", xlab="x: MT scores",
     ylab="y: Calculus Score)",
     main="CS versus MT scores", col.main="darkgreen")
curve(30.18745 + 0.5823085*x, 0,100, add=TRUE, col = "red", lwd = 2)
legend('topleft',inset=0.05,c("y=30.18745+0.5823085x"),
        lty=1,col=c("red"),title="best line fit")
```

The points on the graph appear close to the line of fit. The linear model seems a good description of the relationship between the MT scores and the Calculus scores; students who prepare well for the MT, succeed well in the calculus course.

CS versus MT scores

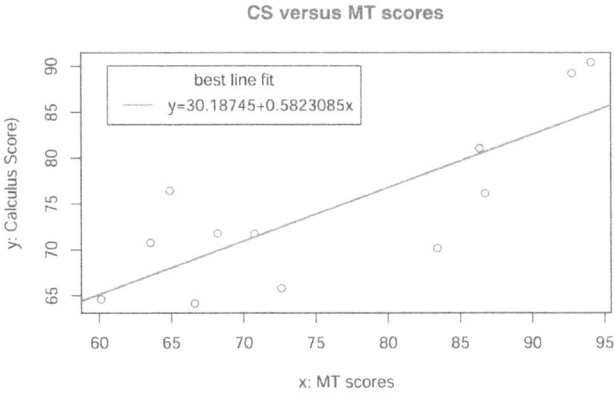

FIGURE 1.1 Calculus scores versus MT scores.

Use of other R codes.

• We apply the "lm" function to y as a function of x, and save the linear regression model in a new variable "fit". Then we read the parameters of the estimated regression equation by calling "fit".

```
fit = lm(y ~ x)
fit
```

```
##
## Call:
## lm(formula = y ~ x)
##
## Coefficients:
## (Intercept)             x
##      30.187         0.582
```

To get the previous plot using "lm", one can use:

```
# plot(x,y,col="blue", main="Calculus Scores versus MT scores",
#                                      abline(lm(y~x)))
```

• To fit y for a particular value of x, we proceed as follows:

```
x=65
y=coefficients(fit)[1] + coefficients(fit)[2]*x
y
```

```
## (Intercept)
##       68.04
```

• Based on the simple linear regression model, if a student obtains a grade of 65 in the MT test, we expect a final score of 68.0375.

Another way to proceed is

```
data=data.frame(x,y)
newdata=data.frame(x=65)       # wrap the parameter
predict(fit, newdata)          # apply predict
```

```
##     1
## 68.04
```

• The "summary" function shows the outputs: residuals, coefficients with related estimates, R^2, ...

```
summary(fit)
```

```
##
## Call:
## lm(formula = y ~ x)
##
## Residuals:
##     Min     1Q Median     3Q    Max
## -8.614 -4.614  0.452  3.938  8.471
##
## Coefficients:
##             Estimate Std. Error t value Pr(>|t|)
## (Intercept)   30.187     10.624    2.84   0.0175 *
## x              0.582      0.138    4.21   0.0018 **
## ---
## Signif. codes:
```

```
## 0 '***' 0.001 '**' 0.01 '*' 0.05 '.' 0.1 ' ' 1
##
## Residual standard error: 5.54 on 10 degrees of freedom
## Multiple R-squared:  0.639,  Adjusted R-squared:  0.603
## F-statistic: 17.7 on 1 and 10 DF,  p-value: 0.00181
```

• To get details on the parameters listed in "summary(lm)", one can use the "help" function:

```
# help(summary.lm)
# help(predict.lm)
```

TABLE 1.2 The first 6 rows of the dataset: cars

speed	dist
4	2
4	10
7	4
7	22
8	16
9	10

1.1.2   Speed and Stopping Distances of Cars

The data set "cars" is a package of R. Recorded in 1920, the data gives the speed of cars and the distances taken to stop. Read the data by tapping: cars, then find a linear representation of the stopping distance versus the speed. Interpret your results.

Solution. First, we read the data (see Table 1.2).

```
top_cars <- head(cars)              # read the first 6 rows of the data
kableExtra::kbl(top_cars, booktabs = TRUE,
            caption = "The first 6 rows of the dataset: cars")
```

```
dim(cars)          # gives the number of rows and columns of the data
```

```
## [1] 50   2
```

```
class(cars)        # type of the data-organization
```

```
## [1] "data.frame"
```

The data contains 50 measurements for each of the numerical variables "speed" and "distance".

Now, we use "ggplot" to plot the data with the best line fit (see Figure 1.2).

```
library(ggplot2)
ggplot(cars,  aes(x=speed, y=dist)) + geom_point() +
        geom_smooth(formula="y ~ x", method=lm, se=F)
```

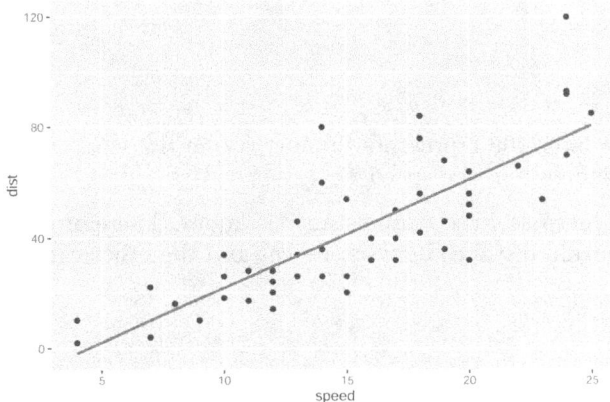

FIGURE 1.2 Stopping distance versus speed

The graph shows a possible linear relationship between the two variables.

Best line fit and comments. Now, we look for the best line fit using the formulas.

```
x=cars$speed
y=cars$dist

sxy=sum((x-mean(x))*(y-mean(y)))
sxx=sum((x-mean(x))*(x-mean(x)))

b = sxy/sxx                         # slope
a = mean(y)- b*mean(x)              # intercept
cat("b = ", b, "      ", "a = ", a)

## b =  3.932       a =  -17.58
```

The equation of the line is: $y = a + bx = -17.57909 + 3.932409x$.

The intercept $a = -17.57909$ represents the predicted value of the stopping distance when the speed is reduced to 0. This value may not be meaningful. However, the line model remains valid within a certain range of the values of x.

The slope $b = 3.932409$ represents the increase in the stopping distance y for every 1-unit increase in the speed x.

As the speed increases, the stop distance is larger. Therefore, a driver has to maintain a certain distance between its car and the other cars for safety.

TABLE 1.3 Inactivity and Hiring Process

Years of Inactivity	Percent of Hospitals Willing to Hire
0.5	100
1.5	94
4.0	75
8.0	44
13.0	28
18.0	17

1.1.3 Returning to Professions after Periods of Inactivity

To better understand the plight of people returning to professions after periods of inactivity, a survey was conducted at some 67 randomly selected hospitals throughout the United States. The administrators were asked about their willingness to hire medical technologists who had been away from the field for a certain number of years. The results are summarized in Table 1.3.

Graph these data and fit them with a straight line. Interpret your results. This is Question 10.4.3 in (R.J. Larsen, 2001).

Solution. Using the "lm" function in R, we can obtain the equation of the best line fit.

```
# Recording Years of Inactivity
x=c(1/2, 3/2, 4, 8, 13, 18)

# Recording Percent of Hospitals Willing to Hire
y=c(100, 94, 75, 44, 28, 17)

# Finding the best line of fit
fit = lm(y ~ x)
fit
```

```
##
## Call:
## lm(formula = y ~ x)
##
## Coefficients:
## (Intercept)              x
##       96.60          -4.92
```

The equation of the line is given by $y = 96.599 - 4.924x$.

There is a decrease of almost 5% of Hospitals willing to hire by an increase of one year of inactivity. The graph shows a line of best fit with a negative slope (see Figure 1.3).

```
library(ggplot2)
mydata=data.frame(x,y)
ggplot(mydata,aes(x=x, y=y))+ geom_point() +
    geom_smooth(formula="y ~ x", method=lm, se=F) +
    labs(x="years of inactivity", y="% of Hospitals willing to hire")
```

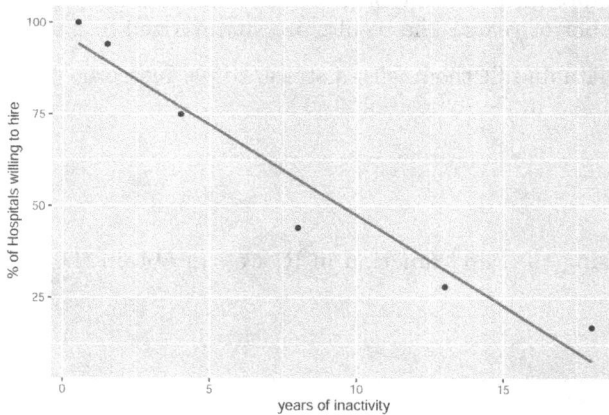

FIGURE 1.3 Hospitals Hiring versus Years of inactivity

1.2   Analysis of Variance for a Linear Regression

Source of variation in y	df: degrees of freedom	SS Sum of Squares	MS = SS/df Mean of SS	F
Regression explained variation	1	$SSR = \dfrac{s_{xy}^2}{s_{xx}}$	$MSR = \dfrac{SSR}{1}$	$F = \dfrac{MSR}{MSE}$
Error unexplained variation	$n-2$	$SSE = s_{yy} - \dfrac{s_{xy}^2}{s_{xx}}$	$MSE = \dfrac{SSE}{n-2} = s^2$	
Total variation in y	$n-1$	$SST = s_{yy}$		

The variation in the response variable y is due to random factors. It is measured by the sample variance of the random variable Y:

$$V(Y) = \frac{1}{n-1}\sum (y_i - \bar{y})^2.$$

When most of the variation in y is due to the linear influence of x, we can obtain a meaningful regression equation. Therefore, it is important to know how much variation in y is due to the linear influence of x and how much is attributed to random factors (Panik, 2012).

From the equality

$$y_i - \bar{y} = y_i - \widehat{y}_i + \widehat{y}_i - \bar{y} = e_i + (\widehat{y}_i - \bar{y})$$

and the fact that $\sum e_i(\widehat{y}_i - \bar{y}) = 0$, one can write:

$$SST = \sum (y_i - \bar{y})^2 = \sum e_i^2 + \sum (\widehat{y}_i - \bar{y})^2 = SSE + SSR$$

Thus, the **total sum of squares** SST of the variation in the response variable y is equal to the **sum of squares of errors** SSE (which explains the variation in y due to random factors), and the **sum of squares of regression** SSR (which reflects the variation of y due to the linear influence of x). These sums are calculated by formulas and organized in the ANOVA table.

The Coefficient of Determination R^2

The ratio

$$R^2 = \frac{SSR}{SST} = \frac{\text{Explained SS}}{\text{Total SS}}; \qquad 0 \leqslant R^2 \leqslant 1,$$

serves as a measure of "goodness of fit"; it represents the proportion of variation in y that can be explained by the linear influence of the variable x. The value of R^2 gets closer to 1 as the relationship gets stronger.

The Analysis of Variance F-test

To test the contribution of the predictor variable x in the regression model better than the simple predictor $\overline{y}$, we perform the F test with the hypotheses:

$$H_0: \quad \beta_1 = 0 \qquad \text{versus} \qquad H_a: \quad \beta_1 \neq 0.$$

The test statistic is found in the ANOVA table as:

$$F = \frac{MSR}{MSE} \quad \rightsquigarrow \quad \mathcal{F}_{df_1=1 \; ; \; df_2=n-2} \quad \text{distributed}$$

Rejection region. With a level of significance α, the null Hypothesis H_0 is rejected if $F > F_\alpha$.

$$\alpha = P\Big(H_0 \text{ is rejected} \,\Big|\, H_0 \text{ is true} \Big) = P\Big(F > F_\alpha \Big).$$

Solved Problems

1.2.1   Points in the Plan

Six points have these coordinates

x	1	2	3	4	5	6
y	5.6	4.6	4.5	3.7	3.2	2.7

1. Find and plot the least squares line for the data.
2. Use the least squares line to predict the value of y when $x = 3.5$.
3. Give an analysis of variance for the linear regression.

Solution. 1. The regression line of best fit is given by $y = a + bx$ where a and b are defined as follows:

$$a = \bar{y} - b\bar{x} \qquad \text{and} \qquad b = \frac{s_{xy}}{s_{xx}}.$$

– A manual calculation gives:

$$\sum x_i = 21 \qquad \sum y_i = 24.3 \qquad \sum x_i y_i = 75.3 \qquad \sum x_i^2 = 91$$

$$\sum y_i^2 = 103.99 \qquad s_{xx} = \sum x_i^2 - \frac{\sum x_i}{n} = 91 - \frac{21^2}{6} = 17.5$$

$$s_{xy} = \sum x_i y_i - \frac{(\sum x_i)(\sum y_i)}{n} = 75.3 - \frac{(21)(24.3)}{6} = 75.3 - 85.05 = -9.75$$

$$b = \frac{s_{xy}}{s_{xx}} = \frac{-9.75}{17.5} = -0.55714, \qquad a = \bar{y} - b\bar{x} = \frac{24.3}{6} - (-0.55714)\left(\frac{21}{6}\right) = 6.$$

– A calculation using R code gives:

```
x=c(1, 2, 3, 4, 5, 6)
y=c(5.6, 4.6, 4.5, 3.7, 3.2, 2.7)

sxy=sum((x-mean(x))*(y-mean(y)))
sxx=sum((x-mean(x))*(x-mean(x)))

b = sxy/sxx
a = mean(y)- b*mean(x)
cat("b = ", b, "        ", "a = ", a)
```

```
## b =   -0.5571           a =  6
```

The equation of the line is: $y = a + bx = 6 - 0.5571429x$.

The slope $b = -0.5571429$ is the amount of decrease in the variable y for every one unit increase in the x coordinate. The intercept is $a = 6$.

2. The line appears to provide a good fit to the data points (see Figure 1.4). When $x = 3.5$, the 2nd coordinate is approximately:

```
x0= 3.5
y0= 6 - 0.5571429*x0
y0
```

```
## [1] 4.05
```

```
plot(x,y,  col = "blue", xlab="x-coordinate",
     ylab="y-coordinate",
     main="Plot of the points", col.main="darkgreen")
points(x0,y0, col="black", pch=19 )
curve( 6 - 0.5571429*x, 0,100, add=TRUE, col = "red", lwd = 2)
legend('topright',inset=0.05,c("y=6-0.5571429x"),
       lty=1,col=c("red"),title="best line fit")
```

3. The analysis of variance for the linear regression is given in the table:

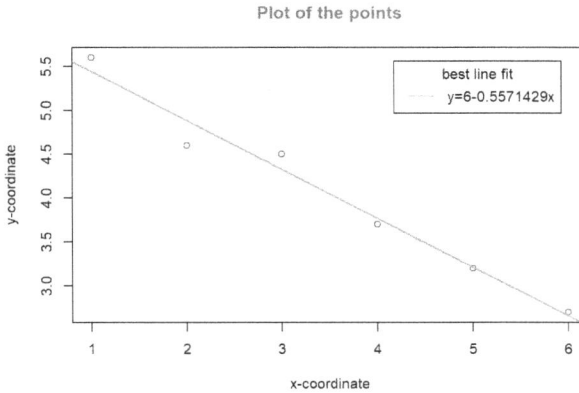

FIGURE 1.4 Points in the plan

Source	df	SS	MS = SS/df
Regression	1	$SSR = \dfrac{s_{xy}^2}{s_{xx}} = \dfrac{(-9.75)^2}{17.5} = 5.4321$	$MSR = \dfrac{SSR}{1} = 5.4321$
Error	$n-2 = 4$	$SSE = s_{yy} - \dfrac{s_{xy}^2}{s_{xx}} = 0.1429$	$MSE = \dfrac{SSE}{n-2} = 0.0357$
Total	$n-1 = 5$	$SST = s_{yy} = SSR + SSE = 5.5750$	

Using R codes, we can form the ANOVA table as follows:

```
x=c(1, 2, 3, 4, 5, 6)
y=c(5.6, 4.6, 4.5, 3.7, 3.2, 2.7)
n=length(x)
syy=sum((y-mean(y))*(y-mean(y)))

SSR=sxy*sxy/sxx
MSR=SSR
SSE=syy-SSR
MSE=SSE/(n-2)

#ANOVA Table
```

```
V1=c("Regression", "Error")
V2=c(1,n-2)
V3=c(SSR, SSE)
V4=c(MSR,MSE)
ANOVA=data.frame(Source=V1, df=V2, SS=V3, MS=V4)
```

```
ANOVA
```

```
##        Source df     SS      MS
## 1 Regression  1 5.4321 5.43214
## 2      Error  4 0.1429 0.03571
```

```
R2=SSR/(SSR+SSE)
R2
```

```
## [1] 0.9744
```

The ratio R^2 shows that 97.43% of the variation in y is explained by the linear model.

TABLE 1.4 MT and CS scores for a larger data set

MT	CS	MT	CS	MT	CS	MT	CS
65.86	66.63	54.76	77.93	24.00	35.24	64.56	79.91
74.48	76.93	48.58	66.32	88.06	90.20	69.42	71.66
74.92	85.43	54.68	65.13	88.34	91.48	58.02	71.68
54.16	53.92	77.00	76.51	52.10	38.05	57.90	59.91
88.94	90.70	45.56	67.23	63.70	76.36	78.08	84.56
80.08	87.05	30.74	42.25	57.60	74.88	NA	NA
34.58	69.64	53.30	72.32	70.58	84.58	NA	NA

1.2.2 ANOVA for MT-CS Framework

i) Construct the ANOVA table for the linear regression of MT-CS "Motivational Test and Calculus Scores" data recorded in Table 1.1.

ii) Find and plot the least squares line for the MT-CS data obtained in a larger class (see Table 1.4). Construct the ANOVA table.

Solution. Let x and y be the vectors of MT scores and CS, respectively. The ANOVA table records the values of the sums of squares of differences between the data y_i, the average $\bar{y}$, and the approximations $\hat{y}_i$. The sums are defined by:

$$SST = s_{yy} = \sum (y_i - \bar{y})^2, \qquad SSR = \sum (\hat{y}_i - \bar{y})^2 = \frac{s_{xy}^2}{s_{xx}},$$

$$SSE = \sum (y_i - \hat{y}_i)^2 = SST - SSR.$$

with

$$s_{xy} = \sum_{i=1}^{n} (x_i - \bar{x})(y_i - \bar{y}), \qquad s_{xx} = \sum_{i=1}^{n} (x_i - \bar{x})^2.$$

i) **ANOVA** table for the 1st data set.

```
n=12
x=c(92.76, 64.90, 72.64, 60.12, 86.72, 66.62, 94.08,
            70.76, 63.56, 86.34, 83.42, 68.20)
y=c(89.24, 76.45, 65.79, 64.58, 76.15, 64.13, 90.44,
            71.71, 70.77, 81.05, 70.15, 71.76)

sxy=sum((x-mean(x))*(y-mean(y)))
sxx=sum((x-mean(x))*(x-mean(x)))
syy=sum((y-mean(y))*(y-mean(y)))

SSR=sxy*sxy/sxx
MSR=SSR
SSE=syy-SSR
MSE=SSE/(n-2)


#Forming the ANOVA Table

V1=c("Regression", "Error")
V2=c(1,n-2)
V3=c(SSR, SSE)
V4=c(MSR,MSE)

ANOVA=data.frame(Source=V1, df=V2,  SS=V3,   MS=V4 )
ANOVA

##         Source df    SS      MS
## 1 Regression   1 543.2 543.24
## 2       Error 10 307.2  30.72

R2=SSR/(SSR+SSE)
R2

## [1] 0.6388

1-R2

## [1] 0.3612
```

The ratio R^2 shows that 63.87% of the variation in y is explained by the linear influence of x; about 36% is unexplained or attributed to random factors.

ii) **ANOVA** table for the 2nd data set

```
n=26
x= c(65.86,74.48,74.92,54.16,88.94,80.08,34.58,54.76,48.58,
     54.68,77.00,45.56,30.74,53.30,24.00,88.06,88.34,52.10,
     63.70,57.60,70.58,64.56,69.42,58.02,57.90,78.08)

y=c(66.63,76.93,85.43,53.92,90.70,87.05,69.64,77.93,66.32,
    65.13,76.51,67.23,42.25,72.32,35.24,90.20,91.48,38.05,
    76.36,74.88,84.58,79.91,71.66,71.68,59.91,84.56)

sxy=sum((x-mean(x))*(y-mean(y)))
sxx=sum((x-mean(x))*(x-mean(x)))
syy=sum((y-mean(y))*(y-mean(y)))

SSR=sxy*sxy/sxx
MSR=SSR
SSE=syy-SSR
MSE=SSE/(n-2)


#Forming the ANOVA Table

V1=c("Regression", "Error")
V2=c(1,n-2)
V3=c(SSR, SSE)
V4=c(MSR,MSE)

ANOVA=data.frame(Source=V1, df=V2,  SS=V3,  MS=V4 )
ANOVA

##          Source df   SS      MS
## 1 Regression   1 4007 4007.13
## 2        Error 24 1936   80.65

library(ggplot2)
mydata=data.frame(x,y)
ggplot(mydata, aes(x=x, y=y))+ geom_point()   +
  geom_smooth(formula="y ~ x", method=lm, se=F) + labs(x="MT", y="CS")
```

The graph shows a linear relationship tendency (see Figure 1.5). The equation of the line is given by:

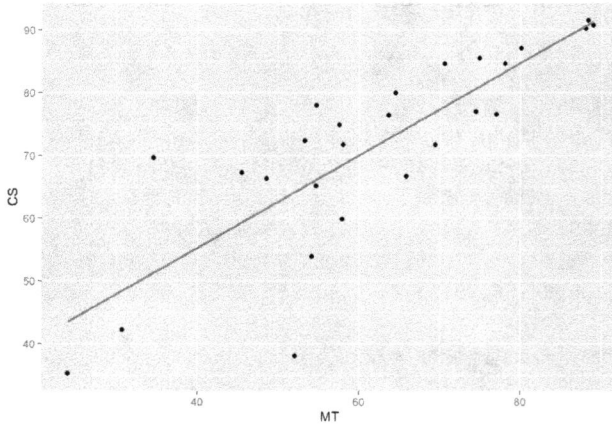

FIGURE 1.5 Calculus scores versus MT scores for a larger data set

```
fit = lm(y ~ x)
fit
```

```
##
## Call:
## lm(formula = y ~ x)
##
## Coefficients:
## (Intercept)            x
##      25.840        0.736
```

There is an increase of 0.7358 in the calculus final scores for every 1-unit increase in the MT score.

Other R codes

The "ANOVA(fit)" function provides a shorter way to read the ANOVA table.

```
anova(fit)
```

```
## Analysis of Variance Table
##
## Response: y
```

```
##              Df Sum Sq Mean Sq F value  Pr(>F)
## x             1   4007    4007    49.7 2.7e-07 ***
## Residuals 24   1936      81
## ---
## Signif. codes:
## 0 '***' 0.001 '**' 0.01 '*' 0.05 '.' 0.1 ' ' 1
```

From the ANOVA table, one can read the values $SSR = 4007.1$ and $SSE = 1935.5$ and deduce the value of the ratio

$$R^2 = \frac{SSR}{SST} = \frac{SSR}{SSR + SSE}.$$

```
R2=4007.1/(4007.1 + 1935.5)
R2
```

```
## [1] 0.6743
```

```
1-R2
```

```
## [1] 0.3257
```

The ratio R^2 shows that 67.42% of the variation in y is explained by the linear influence of x; about 32.5% is unexplained or attributed to random factors. The F testing of the null Hypothesis $H_0 : \beta_1 = 0$ versus the alternative Hypothesis $H_a : \beta_1 \neq 0$, shows the usefulness of the model since the p-value $(= 2.748 \times 10^{-07})$ is very small. Thus, one can consider the introduction of a Motivational Test as a pedagogical tool in order to improve Calculus Scores.

1.3   Correlation Analysis

Correlation Coefficient

Suppose X and Y are two random variables following a "joint bivariate distribution". The covariance between X and Y, defined by

$$cov(X,Y) = E((X - \mu_X)(Y - \mu_Y)) = E(XY) - \mu_X\mu_Y$$

$$\text{with} \qquad \mu_X = E(X), \qquad \mu_Y = E(Y),$$

depicts the joint variation between X and Y.

The population correlation coefficient between X and Y is the covariance of the standardized variables for X and Y:

$$\rho(X,Y) = cov\left(\frac{X - \mu_X}{\sigma_X}, \frac{Y - \mu_Y}{\sigma_Y}\right) = \frac{cov(X,Y)}{\sigma_X\sigma_Y} = \frac{E(XY) - E(X).E(Y)}{\sqrt{Var(X)}\sqrt{Var(Y)}}.$$

We show that

$$|\rho(X,Y)| \leqslant 1 \qquad \text{and}$$

$$|\rho(X,Y)| = 1 \qquad \Longleftrightarrow \qquad Y = mX + p \qquad \begin{array}{c} \text{except possibly} \\ \text{on a set of probability zero.} \end{array}$$

Thus, ρ depicts the strength of the "linear relationship" between X and Y. In this regard (Panik, 2012):

  – For $\rho = 1$ (resp. -1), we have perfect positive (resp. negative) association.

  – If $\rho = 0$, the variables are uncorrelated. This indicates the absence of a
linear relationship between X and Y, but a highly nonlinear relationship may
exist.

  – Correlation does not allow us to predict values of Y from values of X
or vice versa.

Estimating the Correlation Coefficient

Suppose we have a sample of pairs $(x_i, y_i)_{i=1,\ldots,n}$ from the bivariate population
of X and Y values. By estimating each component of ρ by its corresponding
sample moment, we obtain the sample correlation coefficient

$$R = \frac{\frac{1}{n}\sum_{i=1}^{n} x_i y_i - \bar{x}\cdot\bar{y}}{\sqrt{\frac{1}{n}\sum_{i=1}^{n}(x_i - \bar{x})^2}\sqrt{\frac{1}{n}\sum_{i=1}^{n}(y_i - \bar{y})^2}} = \frac{s_{xy}}{\sqrt{s_{xx}s_{yy}}}.$$

Note that the slope in the linear model is given by

$$b = \frac{s_{xy}}{s_{xx}},$$

then

  – If $R = 1$ (resp. -1), we have perfect positive (resp. negative) linear
association.

  – If $R = 0$, the variables are not linearly related but may have a high
nonlinear relationship.

  – $|R| \in (0,1)$ indicates the existence of a linear relationship at a lesser
degree.

Inference on the Correlation Coefficient

Population is bivariate normally distributed	$X \rightsquigarrow \mathcal{N}(\mu_X, \sigma_X^2) \quad Y \rightsquigarrow \mathcal{N}(\mu_Y, \sigma_Y^2) \quad \rho(X,Y)?$
$\downarrow$	
draw (with replacement) random samples of size n	$\textbf{Statistic:} \quad R = \dfrac{S_{xy}}{\sqrt{S_{xx}S_{yy}}}$
$\downarrow$	$R \ \text{is}\ \mathcal{N}\left(E(R)=0, V(R)=\sqrt{\dfrac{1-\rho^2}{n-2}}\right)$
$(x_1,y_1),(x_2,y_2),\ldots,(x_n,y_n)$	$\dfrac{R-E(R)}{V(R)} = R\sqrt{\dfrac{n-2}{1-\rho^2}} \quad \text{is}\ \mathcal{N}(0,1)$

Level of significance.

$$\alpha = P\Big(H_0 \text{ is rejected} \,\Big|\, H_0 \text{ is true}\Big) = P\Big((x,y) \in C \,\Big|\, H_0 \text{ is true}\Big)$$

Test statistic. Let $(x_1,y_1),(x_2,y_2),\ldots,(x_n,y_n)$ be an observed sample value. If H_0 is true, then the statistic

$$t = r\sqrt{\frac{n-2}{1-r^2}} = \frac{\widehat{b}-0}{\sqrt{s_{xy}^2/s_{xx}}} \quad \rightsquigarrow \quad \text{has a } \mathcal{T}_{n-2} \text{ distribution}$$

$$r = \frac{s_{xy}}{\sqrt{s_{xx}s_{yy}}} \qquad\qquad \widehat{b} = \frac{s_{xy}}{s_{xx}}$$

$$s_{xx} = \sum_{i=1}^{n}(x_i - \overline{x})^2 \qquad s_{yy} = \sum_{i=1}^{n}(y_i - \overline{y})^2 \qquad s_{xy} = \sum_{i=1}^{n}(x_i - \overline{x})(y_i - \overline{y})$$

Null Hypothesis	Alternative Hypothesis	Test	Rejection Region C
$H_0:$ $\begin{array}{c}\rho=0\\ \text{or}\\ \rho\leqslant 0\end{array}$	$H_a:\rho>0$	$P(T>t_{\alpha,n-2})=\alpha$ "right sided"	$\{(x,y):\ t>t_{\alpha,n-2}\}$
$H_0:$ $\begin{array}{c}\rho=0\\ \text{or}\\ \rho\geqslant 0\end{array}$	$H_a:\rho<0$	$P(T<-t_{\alpha,n-2})=\alpha$ "left sided"	$\{(x,y):\ t<-t_{\alpha,n-2}\}$
$H_0:\rho=0$	$H_a:\rho\neq 0$	$P(|T|>t_{\frac{\alpha}{2},n-2})=\alpha$ "two sided"	$\{(x,y):\ |t|>t_{\frac{\alpha}{2},n-2}\}$

Confidence Interval (C.I.). The random C.I. (Larson, 1982) is characterized by

$$P\left(-t_{\alpha/2,n-2}<R\sqrt{\frac{n-2}{1-R^2}}<t_{\alpha/2,n-2}\right)=1-\alpha\ .$$

Thus

$$C.I.:\quad -t_{\alpha/2,n-2}\ <\ R\sqrt{\frac{n-2}{1-R^2}}\ <\ t_{\alpha/2,n-2}.$$

It is difficult to use this interval if the inference on the power of the correlation is to be made (R.J. Larsen, 2001).

Direct Inference on the Correlation Coefficient

It would seem natural to use R as a statistic to test more general hypotheses about the correlation ρ, but the probability distribution for R is difficult to obtain. However, the statistic Z_R is near-normally distributed (S. Dowdy, 2004).

Population is bivariate normally distributed	$X \rightsquigarrow \mathcal{N}(\mu_X, \sigma_X^2) \quad Y \rightsquigarrow \mathcal{N}(\mu_Y, \sigma_Y^2) \quad \rho(X,Y)?$
$\downarrow$	
draw (with replacement) random samples of size n	$\rho(X,Y) = \dfrac{E(XY) - E(X).E(Y)}{\sqrt{Var(X)}\sqrt{Var(Y)}} \quad R = \dfrac{S_{xy}}{\sqrt{S_{xx}S_{yy}}}$
$\downarrow$	**Statistic**
$(x_1,y_1),(x_2,y_2),\ldots,(x_n,y_n)$	$Z_R = \dfrac{1}{2}\ln\left(\dfrac{1+R}{1-R}\right)$ is near-normally distributed

Level of Significance.

$$\alpha = P\Big(H_0 \text{ is rejected }\Big|H_0 \text{ is true }\Big) = P\Big((x,y) \in C\Big|H_0 \text{ is true }\Big).$$

Test Statistic.

$$Z = \frac{Z_R - Z_{\rho_0}}{1/\sqrt{n-3}} \quad\rightsquigarrow\quad \text{is approximately } \mathcal{N}(0,1) \text{ distributed}$$

$$E(Z_R) = \frac{1}{2}\ln\left(\frac{1+\rho}{1-\rho}\right), \qquad\qquad V(Z_R) \approx \frac{1}{n-3}.$$

Null Hypothesis	Alternative Hypothesis	Test	Rejection Region C
$H_0:$ $\rho = \rho_0$ or $\rho \leqslant \rho_0$	$H_a : \rho > \rho_0$	$P(Z > z_\alpha) = \alpha$ "right sided"	$\{(x,y):\ z > z_\alpha\}$
$H_0:$ $\rho = \rho_0$ or $\rho \geqslant \rho_0$	$H_a : \rho < \rho_0$	$P(Z < -z_\alpha) = \alpha$ "left sided"	$\{(x,y):\ z < -z_\alpha\}$
$H_0 : \rho = \rho_0$	$H_a : \rho \neq \rho_0$	$P(|Z| > z_{\frac{\alpha}{2}}) = \alpha$ "two sided"	$\{(x,y):\ |z| > z_{\frac{\alpha}{2}}\}$

Decision Rule. Reject H_0 if the observed value z is in the rejection region C. With the decision rule, we achieve a level of significance of α. That is $\alpha\%$ is the mistake made of rejecting H_0 when H_0 is actually true.

Confidence Interval. The $(1-\alpha)100\%$ confidence interval for ρ is

$$z_{\rho_0} - z_{\alpha/2}\left(\frac{1}{\sqrt{n-3}}\right) < z_\rho = \frac{1}{2}\ln\left(\frac{1+\rho}{1-\rho}\right) < z_{\rho_0} + z_{\alpha/2}\left(\frac{1}{\sqrt{n-3}}\right)$$

Inference on the difference of Correlation Coefficients $\rho_1 - \rho_2$

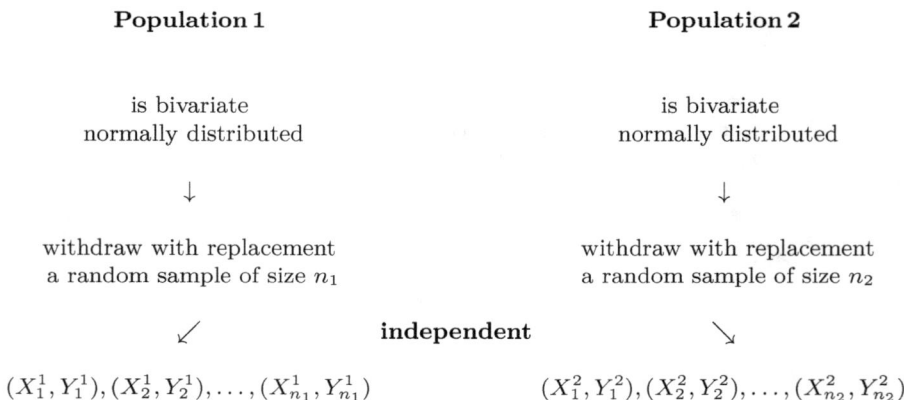

Population 1	**Population 2**
is bivariate normally distributed	is bivariate normally distributed
$\downarrow$	$\downarrow$
withdraw with replacement a random sample of size n_1	withdraw with replacement a random sample of size n_2

$$\swarrow \qquad \textbf{independent} \qquad \searrow$$

$$(X_1^1, Y_1^1), (X_2^1, Y_2^1), \ldots, (X_{n_1}^1, Y_{n_1}^1) \qquad\qquad (X_1^2, Y_1^2), (X_2^2, Y_2^2), \ldots, (X_{n_2}^2, Y_{n_2}^2)$$

Statistics	**Statistics**
$R_1 = \dfrac{S_{x^1 y^1}}{\sqrt{S_{x^1 x^1} S_{y^1 y^1}}}$	$R_2 = \dfrac{S_{x^2 y^2}}{\sqrt{S_{x^2 x^2} S_{y^2 y^2}}}$
$Z_{R_1} = \dfrac{1}{2} \ln\left(\dfrac{1+R_1}{1-R_1}\right)$	$Z_{R_2} = \dfrac{1}{2} \ln\left(\dfrac{1+R_2}{1-R_2}\right)$

Test Statistic. (S. Dowdy, 2004)

$$Z = \frac{Z_{R_1} - Z_{R_2}}{\sqrt{\dfrac{1}{n_1-3} + \dfrac{1}{n_2-3}}} \qquad \text{is approximately } \mathcal{N}(0,1) \text{ for large } n_1, n_2.$$

Null Hypothesis	Alternative Hypothesis	Test	Rejection Region C
$H_0:$ $\rho_1 = \rho_2$ or $\rho_1 \leqslant \rho_2$	$H_a : \rho_1 > \rho_2$	$P(Z > z_\alpha) = \alpha$ "right sided"	$\{(x,y): \ z > z_\alpha\}$
$H_0:$ $\rho_1 = \rho_2$ or $\rho_1 \geqslant \rho_2$	$H_a : \rho_1 < \rho_2$	$P(Z < -z_\alpha) = \alpha$ "left sided"	$\{(x,y): \ z < -z_\alpha\}$
$H_0 : \rho_1 = \rho_2$	$H_a : \rho_1 \neq \rho_2$	$P(|Z| > z_{\frac{\alpha}{2}}) = \alpha$ "two sided"	$\{(x,y): \ |z| > z_{\frac{\alpha}{2}}\}$

Decision Rule. Reject H_0 if the observed value z is in the rejection region C. With the decision rule, we achieve a level of significance of α. That is $\alpha\%$ is the mistake made of rejecting H_0 when H_0 is actually true.

Confidence Interval. The $(1-\alpha)100\%$ confidence interval for $\rho_1 - \rho_2$ is given by:

$$z_{R_1} - z_{R_2} - z_{\alpha/2}\sqrt{\frac{1}{n_1-3} + \frac{1}{n_2-3}}$$
$$< \frac{1}{2}\ln\left(\frac{1+\rho_1}{1-\rho_1}\right) - \frac{1}{2}\ln\left(\frac{1+\rho_2}{1-\rho_2}\right) < z_{R_1} - z_{R_2} + z_{\alpha/2}\sqrt{\frac{1}{n_1-3} + \frac{1}{n_2-3}}$$

Difference between Regression and Correlation

Regression **Model**	Correlation **Model**
1. x is a nonrandom variable, y_x is the expected response of a random variable Y.	**1.** (x, y) is an observed value of a random bivariate variable (X, Y).
2. x is measured under complete control. Only Y contains sampling variability.	**2.** Both X and Y contain sampling variability.
3. For each value of x there is a normal distribution of Y_x.	**3.** For each value of x_i there is a normal distribution of Y_{x_i}, and for each value of y_j there is a normal distribution of X_{y_j}.
4. Each distribution of Y_x has the same variance.	**4.** The X_{y_j} (resp. Y_{x_i}) distributions have the same variance.
5. Linear model: $Y = a + bx + \epsilon$	**5.** Linear model: $Y = a + bx + \epsilon$
6. Expected mean: $E(Y) = a + bx + \epsilon$	**6.** Conditional expectation of Y for a fixed value of X : $E(Y_{|X=x}) = a + bx$
7. $cov(X, Y)$ does not exist and ρ^2 cannot serve as a measure of "covariability".	**7.** X and Y are random variables that follow a joint bivariate distribution. ρ^2 explains the "degree of covariability" between the random variables X and Y.

Regression **Model**	Correlation **Model**
8. $R^2 = \dfrac{SSR}{SST} = \hat{b}\dfrac{s_{xy}}{s_{yy}}$ measures the "goodness of fit" of the population regression line to the scatter of sample points.	**8.** If (X, Y) has a bivariate normal distribution, then $E(Y_{\|X=x}) = \mu_Y + \dfrac{\rho(X,Y)\sigma_X}{\sigma_Y}(x - \mu_X).$
9. The distribution of R depends only on the distribution of Y.	**9.** The distribution of R depends on the distributions of X and Y.
10. $H_0 : b = 0$ tests for the linearity of the model.	**10.** $H_0 : \rho = 0$ or $b = \frac{\sigma_X}{\sigma_Y}\rho = 0$ tests for the independence of X and Y.

Remark. *By convention, when the term "correlation" is seen, it is assumed that Pearson's procedure (for bivariate normal data) is the one under discussion.*

Solved Problems

1.3.1   MT-CS Correlation

The (MT, CS) scores in a Calculus class are recorded in Table 1.1.

1. Based on the graph of the (MT, CS) scores (see Figure 1.1), what will be the sign of the sample correlation coefficient R?

2. Calculate R and R^2 and interpret their values.

3. Is there sufficient evidence to indicate that there is a correlation between these two variables? Test at the 5% level of significance.

Solution.

1. Based on the graph in Figure 1.1, the points are showing a pattern that suggests a positive relationship.

2. The correlation coefficient $\rho = \dfrac{E(XY) - E(X).E(Y)}{\sqrt{Var(X)}\sqrt{Var(Y)}}$ measures the relationship between two variables X and Y when both variables are random. The sign of ρ indicates the direction of the relationship:

- ρ near 0 indicates non-linear relationship

- ρ near +1 or -1 indicates a strong linear relationship.

A point estimator for ρ is:

$$R = \frac{S_{xy}}{\sqrt{S_{xx}S_{yy}}}, \qquad\qquad R^2 = \frac{S_{xy}^2}{S_{xx}S_{yy}}.$$

```
# MT (resp. CS) values are listed in the x (resp. y) vector.
x=c(92.76, 64.90,72.64 ,60.12 ,86.72, 66.62, 94.08, 70.76,
                              63.56, 86.34, 83.42, 68.20)
```

```
y=c(89.24, 76.45, 65.79, 64.58, 76.15, 64.13, 90.44, 71.71,
                                70.77, 81.05, 70.15, 71.76)

sxy=sum((x-mean(x))*(y-mean(y)))
sxx=sum((x-mean(x))*(x-mean(x)))
syy=sum((y-mean(y))*(y-mean(y)))

R2=sxy*sxy/(sxx*syy)
R2
```

```
## [1] 0.6388
```

We have $R^2 = 63.87\%$; that is, 63.87% of the total sum of squares of deviations was reduced by using the least squares equation instead of $\bar{y}$ as a predictor of y.

About 63.87% of the variation in one of the variables is explained by the other.

3. To determine whether there is a significant relationship between the MT scores and the Calculus scores, we test the hypothesis

$$H_0 : \rho = 0 \qquad \text{versus} \qquad H_1 : \rho \neq 0$$

Critical Value Approach: The suitable statistic for the test is:

$$T = \sqrt{\frac{(n-2)R^2}{1-R^2}}, \qquad T \quad \text{is} \quad T_{n-2} \quad \text{distributed}$$

- The **Rejection region** C, at $\alpha = 5\%$ level of significance, is characterized by

$$\alpha = P(|T| > t_{\alpha/2}) = 0.05 \quad \text{and} \quad C = \{(x,y) : t < -t_{\alpha/2}\} \cup \{(x,y) : t > t_{\alpha/2}\}$$

- The **t-value** $t_0 = t_{\alpha/2}$ is equal to:

```
n=12
t0=qt(p=0.05/2, df=n-2, lower.tail = FALSE, log.p = FALSE)
t0
```

```
## [1] 2.228
```

- The **observed statistic** if H_0 was true is

```
n=12
R=sxy/sqrt(sxx*syy)

t=R*sqrt((n-2)/(1-R*R))
t
```

```
## [1] 4.205
```

- **Decision.** The observed value $t = 4.205141 > 2.228139 = t_{\alpha/2}$. It does fall in the rejection region. We can reject H_0.

The data presents sufficient evidence to indicate that a strong relationship exists.

p-value approach: Since we have a two sided test, the p-value is given by

$$p\text{-value} = 2 \times P(T > t) = 2 \times (\text{area to the right of the observed value } t).$$

```
pv=2*pt( q=t, df=n-2, lower.tail = FALSE)
pv
```

```
## [1] 0.001814
```

H_0 is rejected at the level $p - value < 0.01$.

The correlation is declared.

Other codes for reading $R^2, \ldots$

We can read values related to the test by using the function "summary":

```
fit = lm(y ~ x)
summary(fit)
```

```
##
## Call:
## lm(formula = y ~ x)
##
## Residuals:
##     Min      1Q Median     3Q     Max
## -8.614 -4.614  0.452  3.938   8.471
##
## Coefficients:
##             Estimate Std. Error t value Pr(>|t|)
## (Intercept)   30.187     10.624    2.84   0.0175 *
## x              0.582      0.138    4.21   0.0018 **
## ---
## Signif. codes:
## 0 '***' 0.001 '**' 0.01 '*' 0.05 '.' 0.1 ' ' 1
##
## Residual standard error: 5.54 on 10 degrees of freedom
## Multiple R-squared:  0.639,  Adjusted R-squared:  0.603
## F-statistic: 17.7 on 1 and 10 DF,  p-value: 0.00181
```

We can extract the coefficient of determination from the r.squared attribute of
"summary(fit)".

```
fit = lm(y ~ x)
summary(fit)$r.squared
```

```
## [1] 0.6388
```

The coefficient of determination of the simple linear regression model for the
data is 0.6387703.

**Note that the p-value for the F-test is equal to the p-value of the
t-test.** This is because we have $T^2 = F$ and

$$P(F > t^2) = P(T^2 > t^2) = P(|T| > |t| = 4.20514) = 2 \times P(T > t).$$

As the p-value of 0.001814 is less than 0.05, we reject the null Hypothesis:
$\rho = 0$. Hence, there is a significant relationship between the variables in the
linear regression model of the data set.

1.3.2 Speed/Stopping Distances Correlation

Test, at the 5% level of significance, the correlation of the speed and the stopping distance in the data set "cars" of R packages.

Solution. The correlation coefficient is given by:

```
x=cars$speed                          # the speed values
y=cars$dist                           # the stopping distance values


sxy=sum((x-mean(x))*(y-mean(y)))
sxx=sum((x-mean(x))*(x-mean(x)))
syy=sum((y-mean(y))*(y-mean(y)))

R2=sxy*sxy/(sxx*syy)                   # the correlation coefficient
R2
```

```
## [1] 0.6511
```

The value $R^2 = 0.6510794$ means that about 65.1% of the variation in one of the variables is explained by the other.

To test the hypothesis

$$H_0 : \rho = 0 \qquad \text{versus} \qquad H_1 : \rho \neq 0$$

the value of the test statistic is

```
n=length(cars$speed)
t=sqrt((n-2)*R2/(1-R2))
t
```

```
## [1] 9.464
```

The Rejection region C, at $\alpha = 5\%$ level of significance, is characterized by

$$\alpha = P(|T| > t_{\alpha/2}) = 0.05 \quad \text{and} \quad C = \{(x, y) : t < -t_{\alpha/2}\} \cup \{(x, y) : t > t_{\alpha/2}\}$$

- The t-value $t_0 = t_{\alpha/2}$ is equal to:

```
t0=qt(p=0.05/2, df=n-2, lower.tail = FALSE, log.p = FALSE)
t0
```

```
## [1] 2.011
```

Since $t = 9.46399 > 2.010635 = t_{\alpha/2}$, H_0 is rejected and the correlation is declared.

1.3.3   Correlation

Test H_0 concerning the population correlation coefficient.

Would the H_0 be accepted or rejected? What does this mean?

$i)$	$H_0 : \rho = 0$	$H_a : \rho \neq 0$	$n = 20$	$r = 0.55$	$\alpha = 0.01$
$ii)$	$H_0 : \rho = 0$	$H_a : \rho > 0$	$n = 18$	$r = 0.43$	$\alpha = 0.05$
$iii)$	$H_0 : \rho = 0.4$	$H_a : \rho \neq 0.4$	$n = 28$	$r = 0.62$	$\alpha = 0.05$

(This is Question 9.4.2 in D.D. Wackerly (2008).)

Solution.

i) We want to test

$$H_0 : \rho = 0 \qquad \text{against} \qquad H_a : \rho \neq 0.$$

We can either use the t-test or the z-test with the statistic

$$Z = \frac{Z_r - Z_{\rho_0}}{1/\sqrt{n-3}} \qquad \text{with} \qquad \rho_0 = 0 \qquad \text{and} \qquad r = 0.55.$$

The value of the test statistic is

```
rH=0                          # the correlation coefficient if H0 was true

zH=(1/2)*log((1+rH)/(1-rH))
r=0.55                        # the observed correlation
zr=(1/2)*log((1+r)/(1-r))

n=20
se=1/(sqrt(n-3))              # the standard error
```

```
z=(zr-zH)/se
z
```

```
## [1] 2.55
```

The Rejection region C, at $\alpha = 1\%$ level of significance, is characterized by

$$\alpha = P(|Z| > z_{\alpha/2}) = 0.01 \qquad \text{and} \qquad C = [z < -z_{\alpha/2}] \cup [z > z_{\alpha/2}]$$

The z-value $z_0 = z_{\alpha/2}$ is equal to:

```
alpha=0.01
z0=-qnorm(alpha/2, lower.tail = TRUE)
z0
```

```
## [1] 2.576
```

Since $|z| < z_{\alpha/2}$, H_0 is not rejected. We conclude that ρ may be 0.

---

ii) We want to test

$$H_0 : \rho = 0 \qquad \text{against} \qquad H_a : \rho > 0.$$

The test is a z-test with statistic

$$Z = \frac{Z_r - Z_{\rho_0}}{1/\sqrt{n-3}} \qquad \text{with} \qquad \rho_0 = 0 \qquad \text{and} \qquad r = 0.43.$$

The value of the test statistic is

```
rH=0                        # the correlation coefficient if H0 was true
```

```
zH=(1/2)*log((1+rH)/(1-rH))
r=0.43                      # the observed correlation
```

```
zr=(1/2)*log((1+r)/(1-r))

n=18
se=1/(sqrt(n-3))

z=(zr-zH)/se
z
```

```
## [1] 1.781
```

The Rejection region C, at the $\alpha = 5\%$ level of significance, is characterized by

$$\alpha = P(Z > z_\alpha) = 0.05 \qquad \text{and} \qquad C = [z > z_\alpha]$$

The z-value $z_0 = z_\alpha$ is equal to:

```
alpha=0.05
z0=-qnorm(alpha, lower.tail = TRUE)
z0
```

```
## [1] 1.645
```

Since $z > z_\alpha$, H_0 is rejected. We conclude that ρ may be > 0.

---

iii) We want to test

$$H_0 : \rho = 0.4 \qquad \text{against} \qquad H_a : \rho \neq 0.4.$$

The test is a z-test with statistic

$$Z = \frac{Z_r - Z_{\rho_0}}{1/\sqrt{n-3}} \qquad \text{with} \qquad \rho_0 = 0.4 \qquad \text{and} \qquad r = 0.62.$$

The value of the test statistic is

```
rH=0.4                      # the correlation coefficient if H0 was true

zH=(1/2)*log((1+rH)/(1-rH))
r=0.62                      # the observed correlation
zr=(1/2)*log((1+r)/(1-r))

n=28
se=1/(sqrt(n-3))

z=(zr-zH)/se
z

## [1] 1.507
```

The Rejection region C, at the $\alpha = 5\%$ level of significance, is characterized by

$$\alpha = P(|Z| > z_{\alpha/2}) = 0.05 \qquad \text{and} \qquad C = [z < -z_{\alpha/2}] \cup [z > z_{\alpha/2}]$$

The z-value $z_0 = z_{\alpha/2}$ is equal to:

```
z0=-qnorm(alpha/2, lower.tail = TRUE)
z0

## [1] 1.96
```

Since $|z| < z_{\alpha/2}$, H_0 is not rejected. We conclude that ρ may be 0.4.

1.3.4 Obesity

i)   In a study of Obesity, the sample correlation for weights of 28 mature obese brother-sister pairs is computed to be $r = 0.64$. A nutritionist wishes to place a 95% confidence interval on the population correlation coefficient ρ. Find this interval.

ii) Because of some prior theory, test for the nutritionist:

$$H_0 : \rho = 0.5 \qquad \text{against} \qquad H_a : \rho \neq 0.5.$$

iii) Suppose that the nutritionist has data on 23 brother-sister pairs of conventional mature weight in addition to the data above for obese pairs where $r_1 = 0.64$. For the conventional sample, $r_2 = 0.38$. Test whether the correlation is the same for both populations at $\alpha = 0.05$.

(These are questions formulated from examples 9.3, 9.4, 9.5 in D.D. Wackerly (2008).)

Solution.

i) For n large, the statistic:

$$Z = \frac{Z_r - Z_\rho}{1/\sqrt{n-3}} \quad \text{is approximatively } N(0,1) \quad \text{with} \quad Z_r = \frac{1}{2}\ln\left(\frac{1+r}{1-r}\right)$$

and r is the sample correlation coefficient.

A 95% confidence interval on the population correlation coefficient ρ is first obtained on the transformed parameter z_ρ using $r = 0.64$:

$$z_r - z_{\alpha/2}\left(\frac{1}{\sqrt{n-3}}\right) \; < \; z_\rho = \frac{1}{2}\ln\left(\frac{1+\rho}{1-\rho}\right) \; < \; z_r + z_{\alpha/2}\left(\frac{1}{\sqrt{n-3}}\right)$$

```
n=28
alpha=0.05
z0=-qnorm(alpha/2, lower.tail = TRUE)        # z-value
se=1/(sqrt(n-3))                             # standard error
r=0.64
zr=(1/2)*log((1+r)/(1-r))                     # observed value


LCB= zr - z0*se              # Lower confidence bound of the C.I
UCB= zr + z0*se              # Upper confidence bound of the C.I
cat("LCB = ", LCB, "    ", "UCB = ", UCB)


## LCB =  0.3662      UCB =  1.15
```

Thus, the 95% confidence interval for z_ρ is

$$0.3661809 < z_\rho < 1.150167$$

and we deduce the 95% confidence interval for ρ by computing:

```
a=exp(2*LCB)
b=exp(2*UCB)
rLCB = (a-1)/(1+a)            # Lower confidence bound of the C.I
rUCB = (b-1)/(1+b)            # Upper confidence bound of the C.I
cat("rLCB = ", rLCB, "    ", "rUCB = ", rUCB)


## rLCB =  0.3506      rUCB =  0.8178
```

Thus,

$$C.I_{0.95}: \qquad 0.3506467 < \rho < 0.8178092.$$

---

ii) Because of some prior theory or available evidence, the nutritionist wants to test

$$H_0 : \rho = 0.5 \qquad\text{against}\qquad H_a : \rho \neq 0.5.$$

The test is a z-test with statistic

$$Z = \frac{Z_r - Z_{\rho_0}}{1/\sqrt{n-3}} \qquad \text{with} \qquad \rho_0 = 0.5 \qquad \text{and} \qquad r = 0.64.$$

The value of the observed statistic is

```
rH=0.5                          # the correlation coefficient if H0 was true
zH=(1/2)*log((1+rH)/(1-rH))
z=(zr-zH)/se
z
```

```
## [1] 1.044
```

The Rejection region C, at $\alpha = 5\%$ level of significance, is characterized by

$$\alpha = P(|Z| > z_{\alpha/2}) = 0.05 \qquad \text{and}$$
$$C = \{(x,y) : z < -z_{\alpha/2}\} \cup \{(x,y) : z > z_{\alpha/2}\}$$

The z-value $z_0 = z_{\alpha/2}$ is equal to:

```
z0=-qnorm(alpha/2, lower.tail = TRUE)
z0
```

```
## [1] 1.96
```

Since $|z| < z_{\alpha/2}$, H_0 is not rejected. The nutritionist concludes that ρ may be 0.5.

---

iii) The following hypotheses are considered:

$$H_0 : \rho_1 = \rho_2 \qquad \text{against} \qquad H_a : \rho_1 \neq \rho_2.$$

and are tested with the statistic

$$Z = \frac{Z_{R_1} - Z_{R_2}}{\sqrt{\dfrac{1}{n_1 - 3} + \dfrac{1}{n_2 - 3}}} \qquad \text{which is approximately } N(0,1).$$

Thus, the observed statistic is equal to

```
n1=28
n2=23
r1=0.64
r2=0.38
z1=(1/2)*log((1+r1)/(1-r1))
z2=(1/2)*log((1+r2)/(1-r2))

z=(z1-z2)/sqrt(1/(n1-3) + 1/(n2-3))
z
```

```
## [1] 1.194
```

Since $z_{\alpha/2} = 1.96$ and $z = 1.93714 < 1.96 = z_{\alpha/2}$, the null Hypothesis H_0 is not rejected. There is no significant difference between the two correlation coefficients.

The correlation between weights of brother-sister pairs may be the same for obese siblings as for those of non-obese siblings.

1.4   Inference on the Slope and the y-Intercept

Theorem. Assume

$$
\begin{array}{ll}
Y_1, Y_2, \ldots, Y_n \text{ are independent} & Y_i = Y_{x_i} \rightsquigarrow \mathcal{N}(\mu_i, \sigma^2) \quad i = 1, \ldots, n \\[2mm]
\mu_i = E(Y_i) = a + b x_i & Var(Y_i) = \sigma^2 \quad i = 1, \ldots, n \\[2mm]
x_1, \ldots, x_n \text{ are constants} & a, b \text{ are unknown parameters}
\end{array}
$$

Then, we have

$$
\frac{\widehat{A} - a}{\sigma_{\widehat{A}}} \quad \text{and} \quad \frac{\widehat{B} - b}{\sigma_{\widehat{B}}} \quad \rightsquigarrow \quad \mathcal{N}(0,1)
$$

$$
\widehat{A} \ \ (\text{resp. } \widehat{B}) \quad \text{and} \quad \frac{(n-2)S}{\sigma^2} \quad \text{are independent}
$$

$$
\frac{(n-2)S}{\sigma^2} \quad \rightsquigarrow \quad \chi^2_{n-2}
$$

$$
\frac{\widehat{A} - a}{S_{\widehat{A}}} \quad \text{and} \quad \frac{\widehat{B} - b}{S_{\widehat{B}}} \quad \rightsquigarrow \quad \mathcal{T}_{n-2}
$$

$$
\sigma^2_{\widehat{A}} = Var(\widehat{A}) = \frac{\sigma^2}{n} \cdot \frac{\displaystyle\sum_{i=1}^{n} x_i^2}{\displaystyle\sum_{i=1}^{n} (x_i - \overline{x})^2} \qquad\qquad \sigma^2_{\widehat{B}} = Var(\widehat{B}) = \frac{\sigma^2}{\displaystyle\sum_{i=1}^{n} (x_i - \overline{x})^2}
$$

$$
S^2_{\widehat{A}} = \frac{S^2}{n} \cdot \frac{\displaystyle\sum_{i=1}^{n} x_i^2}{\displaystyle\sum_{i=1}^{n} (x_i - \overline{x})^2} \qquad\qquad S^2_{\widehat{B}} = \frac{S^2}{\displaystyle\sum_{i=1}^{n} (x_i - \overline{x})^2}
$$

$$
E(\widehat{A}) = a \qquad\qquad E(\widehat{B}) = b \qquad\qquad cov(\widehat{A}, \widehat{B}) = \frac{-\sigma^2 \overline{x}}{\displaystyle\sum_{i=1}^{n} (x_i - \overline{x})^2}
$$

where the least squares estimators for a and b are

$$\widehat{A} = \overline{Y} - \widehat{B}\overline{x} \qquad\qquad \widehat{B} = \frac{\displaystyle\sum_{i=1}^{n}(x_i - \overline{x})(Y_i - \overline{Y})}{\displaystyle\sum_{i=1}^{n}(x_i - \overline{x})^2}$$

and the estimator for σ^2 is

$$S^2 = \frac{1}{n-2}\left[\sum_{i=1}^{n}(Y_i - \overline{Y})^2 - \widehat{B}^2\sum_{i=1}^{n}(x_i - \overline{x})^2\right]$$

with

$$\overline{x} = \frac{1}{n}\sum_{i=1}^{n}x_i \qquad\qquad \overline{Y} = \frac{1}{n}\sum_{i=1}^{n}Y_i$$

Inference on the slope of the line of means a + bx

Population is normally distributed	$E(Y_i) = a + bx_i \qquad b?$
$\downarrow$ for each $x = (x_1, x_2, \ldots, x_n)$ draw (with replacement) random samples of size n	$Y_1, Y_2, \ldots, Y_n \quad$ independent
$\downarrow$ $y = (y_1, y_2, \ldots, y_n)$ $= (y_{x_1}, y_{x_2}, \ldots, y_{x_n})$	**Statistic** $T = \dfrac{\widehat{B} - b}{S_{\widehat{B}}} \quad\rightsquigarrow\quad T_{n-2}$

Level of significance.

$$\alpha = P\Big(H_0 \text{ is rejected } \Big| H_0 \text{ is true}\Big) = P\Big(y \in C \Big| H_0 \text{ is true}\Big)$$

Test statistic. Let $y = (y_1, y_2, \ldots, y_n)$ be an observed sample value corresponding to $x = (x_1, x_2, \ldots, x_n)$. Set

$$t = \frac{b - b_0}{s_b} \qquad b = \frac{s_{xy}}{s_{xx}} \qquad s_b^2 = \frac{s^2}{s_{xx}} \qquad s^2 = \frac{1}{n-2}\left[s_{yy}^2 - \frac{s_{xy}^2}{s_{xx}}\right]$$

Null Hypothesis	Alternative Hypothesis	Test	Rejection Region C
$H_0:$ $\begin{array}{c} b = b_0 \\ \text{or} \\ b \leqslant b_0 \end{array}$	$H_a : b > b_0$	$P(T > t_{\alpha,n-2}) = \alpha$ right sided	$\{y : \ t > t_{\alpha,n-2}\}$
$H_0:$ $\begin{array}{c} b = b_0 \\ \text{or} \\ b \geqslant b_0 \end{array}$	$H_a : b < b_0$	$P(T < -t_{\alpha,n-2}) = \alpha$ left sided	$\{y : \ t < -t_{\alpha,n-2}\}$
$H_0 : b = b_0$	$H_a : b \neq b_0$	$P(|T| > t_{\frac{\alpha}{2},n-2}) = \alpha$ two sided	$\{y : \ |t| > t_{\frac{\alpha}{2},n-2}\}$

Decision rule.   Reject H_0 if the observed value x is in the rejection region C. With the decision rule, we achieve a level of significance of α. That is, $\alpha\%$ is the mistake made of rejecting H_0 when H_0 is actually true.

Confidence Interval.   The random C.I. is characterized by

$$P\Big(-t_{\alpha/2,n-2} < \frac{\widehat{B} - b_0}{S_{\widehat{B}}} < t_{\alpha/2,n-2}\Big) = 1 - \alpha$$

Solved Problems

1.4.1   MT-CS Linear Relationship

Consider the MT ("Motivational Test") scores and Calculus scores "CS" obtained in a Calculus class (see Table 1.1).

1. Determine whether there is a significant linear relationship between the MT scores and the Calculus scores.

2. Find a 95% confidence interval estimate of the slope b of the least squares line.

Solution. 1. The hypothesis to be tested is:

$$H_0 : b = 0 \qquad\qquad \text{versus} \qquad\qquad H_1 : b \neq 0$$

where b is the slope in the linear model $E(Y_{|x}) = a + bx$. The suitable statistic for the test is:

$$T = \frac{b - 0}{\sqrt{MSE/s_{xx}}}, \qquad\qquad T \quad \text{is} \quad \mathcal{T}_{n-2}$$

The Rejection region C, at the $\alpha = 5\%$ level of significance, is characterized by

$$\alpha = P(|T| > t_{\alpha/2}) = 0.05 \quad \text{and} \quad C = \{y : t < -t_{\alpha/2}\} \cup \{y : t > t_{\alpha/2}\}$$

The t-value $t_0 = t_{\alpha/2}$ is equal to:

```
n=12
t0=qt(p=0.05/2, df=n-2, lower.tail = FALSE, log.p = FALSE)
t0
```

```
## [1] 2.228
```

The observed statistic if H_0 was true is t, calculated as follows:

```
x=c(92.76, 64.90,72.64 ,60.12 ,86.72, 66.62, 94.08, 70.76,
                                 63.56, 86.34, 83.42, 68.20)
y=c(89.24, 76.45, 65.79, 64.58, 76.15, 64.13, 90.44, 71.71,
                                 70.77, 81.05, 70.15, 71.76)

sxy=sum((x-mean(x))*(y-mean(y)))
sxx=sum((x-mean(x))*(x-mean(x)))
syy=sum((y-mean(y))*(y-mean(y)))

SSR=sxy*sxy/sxx
SSE=syy-SSR
MSE=SSE/(n-2)
se=sqrt(MSE/sxx)

b = sxy/sxx
t=(b-0)/(se)
t
```

```
## [1] 4.205
```

The observed value $t = 4.205141 > 2.228139 = t_{\alpha/2}$. It does fall in the rejection region. We can reject H_0.

The data presents sufficient evidence to indicate that a linear relationship exists.

2. The 95% Confidence interval of the slope b for the data is given by:

```
LCB= b - (t0)*se              #  Lower bound of the C.I
UCB= b + (t0)*se              #  Upper bound of the C.I
LCB
```

```
## [1] 0.2738
```

```
UCB
```

```
## [1] 0.8909
```

Since the C.I.: $(0.2737661, 0.8908509)$ doesn't contain 0, we can conclude that the true value of the slope b is not 0. Thus, we can reject H_0 in favor of H_1.

Other R codes for finding a Confidence Interval

We can deduce a *C.I* of *b* by using the information in the "summary":

```
LR = lm(y ~ x)
data=data.frame(x,y)
summary(LR)
```

```
##
## Call:
## lm(formula = y ~ x)
##
## Residuals:
##     Min      1Q Median     3Q    Max
## -8.614 -4.614  0.452  3.938  8.471
##
## Coefficients:
##             Estimate Std. Error t value Pr(>|t|)
## (Intercept)   30.187     10.624    2.84   0.0175 *
## x              0.582      0.138    4.21   0.0018 **
## ---
## Signif. codes:
## 0 '***' 0.001 '**' 0.01 '*' 0.05 '.' 0.1 ' ' 1
##
## Residual standard error: 5.54 on 10 degrees of freedom
## Multiple R-squared:  0.639,  Adjusted R-squared:  0.603
## F-statistic: 17.7 on 1 and 10 DF,  p-value: 0.00181
```

The estimate of the observed slope is the coefficient of x. It is equal to 0.5823 and the standard error is 0.1385. Thus one can deduce the C.I as above.

1.4.2   Speed/Stopping Distances Linear Relationship

Test, at the 5% level of significance, the Linear Relationship of the speed and the stopping distance in the data set "cars" from R packages.

Solution. To test the hypothesis

$$H_0 : b = 0 \qquad \text{versus} \qquad H_1 : b \neq 0$$

in the linear model $E(Y_{|x}) = a + bx$, we calculate the value of the test statistic:

```
x=cars$speed                          # speed values
y=cars$dist                           # stopping distance values
n=length(x)

sxy=sum((x-mean(x))*(y-mean(y)))
sxx=sum((x-mean(x))*(x-mean(x)))
syy=sum((y-mean(y))*(y-mean(y)))

SSR=sxy*sxy/sxx
SSE=syy-SSR
MSE=SSE/(n-2)

se=sqrt(MSE/sxx)

b = sxy/sxx                           # observed slope
t = (b-0)/se                          # observed statistic
t

## [1] 9.464
```

The Rejection region C, at $\alpha = 5\%$ level of significance, is characterized by

$$\alpha = P(|T| > t_{\alpha/2}) = 0.05 \qquad \text{and} \qquad C = \{y : t < -t_{\alpha/2}\} \cup \{y : t > t_{\alpha/2}\}.$$

The t-value $t_0 = t_{\alpha/2}$ is equal to:

```
t0=qt(p=0.05/2, df=n-2, lower.tail = FALSE, log.p = FALSE)
t0
```

```
## [1] 2.011
```

Since $t = 9.46399 > 2.010635 = t_{\alpha/2}$, H_0 is rejected and the strong linear relationship is confirmed.

1.5   Estimation of the Mean $E(Y)$ When $x = x^*$

Population is normally distributed	$E(Y) = a + bx$ $\mathbf{E(Y_{x^*}) = a + bx^*}$?
$\downarrow$	$Y_1, Y_2, \ldots, Y_n \quad$ independent
for each $x = (x_1, x_2, \ldots, x_n)$ draw (with replacement) random samples of size n $\downarrow$	**Statistic**
$y = (y_1, y_2, \ldots, y_n)$	$\widehat{Y}_{x^*} = \widehat{A} + \widehat{B}x^*$
$= (y_{x_1}, y_{x_2}, \ldots, y_{x_n})$	$T = \dfrac{\widehat{Y}_{x^*} - (a + bx^*)}{S_{x^*}} \quad \rightsquigarrow \quad T_{n-2}$

Level of significance.

$$\alpha = P\Big(H_0 \text{ is rejected } \Big| H_0 \text{ is true }\Big) = P\Big(y \in C \Big| H_0 \text{ is true }\Big)$$

Test statistic. Let $y = (y_1, y_2, \ldots, y_n)$ be an observed sample value corresponding to $x = (x_1, x_2, \ldots, x_n)$. Set

$$t = \frac{\widehat{y}_{x^*} - \mu_0}{s_{x^*}} \qquad\qquad \mu = E(Y_{x^*}) = a + bx^* = E(\widehat{Y}_{x^*})$$

$$\widehat{Y}_{x^*} = \widehat{A} + \widehat{B}x^* = (\overline{Y}_{x^*} - \widehat{B}\overline{x}) + \widehat{B}x^* = \overline{Y}_{x^*} + \widehat{B}(x^* - \overline{x}) \qquad \overline{x} = \frac{1}{n}\sum_{i=1}^{n} x_i$$

$$\sigma_{x^*}^2 = Var(\widehat{Y}_{x^*}) = \sigma^2\left[\frac{1}{n} + \frac{(x^* - \overline{x})^2}{s_{xx}}\right] \qquad\qquad s_{xx} = \sum_{i=1}^{n} (x_i - \overline{x})^2$$

$$s_{x^*}^2 = s^2\left[\frac{1}{n} + \frac{(x^* - \overline{x})^2}{s_{xx}}\right] \qquad \text{an estimator of } \sigma_{x^*}^2$$

Null Hypothesis	Alternative Hypothesis	Test	Rejection Region C
$H_0 : \mu \leqslant \mu_0$	$H_a : \mu > \mu_0$	$P(T > t_{\alpha,n-2}) = \alpha$ right sided	$\{y : t > t_{\alpha,n-2}\}$
$H_0 : \mu \geqslant \mu_0$	$H_a : \mu < \mu_0$	$P(T < -t_{\alpha,n-2}) = \alpha$ left sided	$\{y : t < -t_{\alpha,n-2}\}$
$H_0 : \mu = \mu_0$	$H_a : \mu \neq \mu_0$	$P(|T| > t_{\frac{\alpha}{2},n-2}) = \alpha$ two sided	$\{y : |t| > t_{\frac{\alpha}{2},n-2}\}$

Decision Rule. Reject H_0 if the observed value x is in the rejection region C. With the decision rule, we achieve a level of significance of α. That is $\alpha\%$ is the mistake made of rejecting H_0 when H_0 is actually true.

Confidence Interval. The random C.I. is characterized by

$$P\left(-t_{\alpha/2,n-2} < \frac{\widehat{Y}_{x^*} - (a + bx^*)}{S_{x^*}} < t_{\alpha/2,n-2}\right) = 1 - \alpha$$

For $x = x^*$, a $(1 - \alpha)100\%$ confidence interval for $E(Y) = a + bx$ is

$$\widehat{y}_{x^*} - t_{\alpha/2,n-2}\, s_{x^*} < a + bx < \widehat{y}_{x^*} + t_{\alpha/2,n-2}\, s_{x^*}$$

Solved Problems

1.5.1   MT-CS Linear Relationship

The following are MT scores and Calculus scores "CS" obtained in a Calculus class (see Table 1.1).

1. Estimate the average calculus scores for students whose MT score is 50 with a 95% confidence interval.

2. Do the data support the hypothesis of a 0 intercept? Use $\alpha = 0.05$.

Solution.

1. The point estimate of $E(Y|_{x=50})$; the average calculus score for students whose MT score is 50, is calculated as follows:

```
x=c(92.76, 64.90,72.64 ,60.12 ,86.72, 66.62, 94.08, 70.76,
                          63.56, 86.34, 83.42, 68.20)
y=c(89.24, 76.45, 65.79, 64.58, 76.15, 64.13, 90.44, 71.71,
                          70.77, 81.05, 70.15, 71.76)
sxy=sum((x-mean(x))*(y-mean(y)))
sxx=sum((x-mean(x))*(x-mean(x)))
syy=sum((y-mean(y))*(y-mean(y)))

b = sxy/sxx                      # slope
a = mean(y)- b*mean(x)           # intercept
x0=50
y0=a+b*x0                 # the predicted mean using the linear
y0                        # model E(Y|x)=a+bx
```

```
## [1] 59.3
```

The 95% confidence interval for CS average when the MT score is $x = 50$, is calculated as follows:

```
n=length(x)
SSR=sxy*sxy/sxx
SSE=syy-SSR
MSE=SSE/(n-2)


var=MSE*( 1/n + (x0-mean(x))*(x0-mean(x))/sxx )
s0=sqrt(var)


t0=qt(p=0.05/2, df=n-2, lower.tail = FALSE)


LCB= y0 - (t0)*s0                          #  Lower bound of the C.I
UCB= y0 + (t0)*s0                          #  Upper bound of the C.I
cat("LCB = ", LCB, "        ", "UCB = ", UCB)


## LCB =        UCB =  68.04
```

The $C.I$ is $(50.56842, 68.03732)$.

Although, we do not know whether the particular interval $(50.56842, 68.03732)$ contains $E(Y) = a + bx$, the procedure that generates it yields intervals that do capture the mean.

In the long run, 95% of the intervals constructed in this way will contain the unknown mean $E(Y) = a + bx$ when $x = 50$.

Other R codes for reading a Confidence Interval of the mean $E(Y)$

We can read a $C.I$ of $E(Y)$ by using the function "predict":

```
LR = lm(y ~ x)
data=data.frame(x,y)
newdata=data.frame(x=50)                   # wrap the parameter
predict(LR,newdata,level = 0.95,interval="confidence")


##     fit   lwr   upr
## 1 59.3 50.57 68.04
```

2. The hypothesis to be tested is

$$H_0 : \quad \mu = E(y) = 0 \qquad \text{versus} \qquad H_1 : \quad \mu \neq 0$$

The suitable statistic for the test is:

$$T = \frac{m - 0}{\sigma_m}, \qquad\qquad T \quad \text{is} \quad T_{n-2}$$

The Rejection region C, at the $\alpha = 5\%$ level of significance, is characterized by

$$\alpha = P(|T| > t_{\alpha/2}) = 0.05 \quad \text{and} \quad C = \{y : t < -t_{\alpha/2}\} \cup \{y : t > t_{\alpha/2}\}.$$

The t-value $t_0 = t_{\alpha/2}$ is equal to:

```
t0=qt(p=0.05/2, df=n-2, lower.tail = FALSE, log.p = FALSE)
t0
```

```
## [1] 2.228
```

The observed statistic if H_0 was true is

```
x1=0
m1=a+b*x1

var1=MSE*(1/n + (x1-mean(x))*(x1-mean(x))/sxx )
s1=sqrt(var1)

t1=(m1-0)/(s1)
t1
```

```
## [1] 2.842
```

Since $|t_1| < t_{\alpha/2}$, the observed value doesn't fall in the rejection region. Therefore, we do not reject H_0 and support H_1. It is unlikely that the mean Calculus score is zero. We should include a non zero intercept in the model $y = a + bx + \epsilon$.

1.6   Inferences about Future Observations

Population	$E(Y_{x^*}) = a + bx^*$ $\mathbf{Y_{x^*}}$?
is normally distributed	$\widehat{Y}_{x^*} = \widehat{A} + \widehat{B}x^*$

$\downarrow$

for each $x = (x_1, x_2, \ldots, x_n)$ draw	$Y_{x^*}, Y_1, Y_2, \ldots, Y_n$ independent
(with replacement)	y_{x^*}: hypothetical future observation
random samples of size n	

$\downarrow$ Statistic

$$y = (y_1, y_2, \ldots, y_n)$$

$$= (y_{x_1}, y_{x_2}, \ldots, y_{x_n})$$

$$T = \frac{\widehat{Y}_{x^*} - Y_{x^*}}{S_{x^*+e}} \quad \rightsquigarrow \quad \mathcal{T}_{n-2}$$

Level of Significance.

$$\alpha = P\Big(H_0 \text{ is rejected } \Big| H_0 \text{ is true}\Big) = P\Big(y \in C \Big| H_0 \text{ is true}\Big)$$

Test Statistic. Let $y = (y_1, y_2, \ldots, y_n)$ be an observed sample value corresponding to $x = (x_1, x_2, \ldots, x_n)$. We have

$$t = \frac{\widehat{y}_{x^*} - y_0}{s_{x^*+e}} \qquad\qquad E(Y_{x^*}) = a + bx^* = E(\widehat{Y}_{x^*})$$

$$E(Y_{x^*} - \widehat{Y}_{x^*}) = 0 \qquad \overline{x} = \frac{1}{n}\sum_{i=1}^{n} x_i \qquad s_{xx} = \sum_{i=1}^{n} (x_i - \overline{x})^2$$

$$\sigma^2_{x^*+e} = Var(Y_{x^*} - \widehat{Y}_{x^*}) = \sigma^2 + \sigma^2_{x^*} = \sigma^2 \left[1 + \frac{1}{n} + \frac{(x^* - \bar{x})^2}{S_{xx}}\right]$$

$$s^2_{x^*+e} = s^2 \left[1 + \frac{1}{n} + \frac{(x^* - \bar{x})^2}{S_{xx}}\right] \quad \text{an estimator of } \sigma^2_{x^*+e}.$$

Null Hypothesis	Alternative Hypothesis	Test	Rejection Region C
$H_0 : y \leqslant y_0$	$H_a : y > y_0$	$P(T > t_{\alpha,n-2}) = \alpha$ right sided	$\{y :\ t > t_{\alpha,n-2}\}$
$H_0 : y \geqslant y_0$	$H_a : y < y_0$	$P(T < -t_{\alpha,n-2}) = \alpha$ left sided	$\{y :\ t < -t_{\alpha,n-2}\}$
$H_0 : y = y_0$	$H_a : y \neq y_0$	$P(|T| > t_{\frac{\alpha}{2},n-2}) = \alpha$ two sided	$\{y :\ |t| > t_{\frac{\alpha}{2},n-2}\}$

Decision Rule. Reject H_0 if the observed value x is in the rejection region C. With the decision rule, we achieve a level of significance of α. That is, $\alpha\%$ is the mistake made of rejecting H_0 when H_0 is actually true.

Confidence Interval. The random C.I. is characterized by

$$P\left(-t_{\alpha/2,n-2} < \frac{\widehat{Y}_{x^*} - Y_{x^*}}{S_{x^*+e}} < t_{\alpha/2,n-2}\right) = 1 - \alpha$$

For predicting a particular value of y when $x = x^*$, we use an $(1 - \alpha)100\%$ confidence and prediction interval

$$\widehat{y}_{x^*} - t_{\alpha/2,n-2}\, s_{x^*+e} < y_{x^*} < \widehat{y}_{x^*} + t_{\alpha/2,n-2}\, s_{x^*+e}$$

Solved Problems

1.6.1   MT-CS Future Observations

The MT scores and Calculus scores "CS" obtained in a Calculus class are recorded in Table 1.1.

A student took the MT and scored 50 but has not yet taken the calculus test. Predict the calculus score for this student with a 95% confidence interval.

Solution. The predicted value of y is given by:

```
# Mt (resp. CS) scores are stored in x (resp. y)
x=c(92.76, 64.90,72.64 ,60.12 ,86.72, 66.62, 94.08, 70.76,
                            63.56, 86.34, 83.42, 68.20)
y=c(89.24, 76.45, 65.79, 64.58, 76.15, 64.13, 90.44, 71.71,
                            70.77, 81.05, 70.15, 71.76)

sxy=sum((x-mean(x))*(y-mean(y)))
sxx=sum((x-mean(x))*(x-mean(x)))
syy=sum((y-mean(y))*(y-mean(y)))

b = sxy/sxx                      # slope of best line fit
a = mean(y)- b*mean(x)           # intercept

x0=50
y0=a+b*x0                        # predicted value
y0

## [1] 59.3
```

The 95% confidence interval (C.I.) is calculated as follows:

```
n=length(x)                    # sample size
SSR=sxy*sxy/sxx
SSE=syy-SSR
MSE=SSE/(n-2)


var=MSE*(1 + 1/n + (x0-mean(x))*(x0-mean(x))/sxx )

s0=sqrt(var)                         # standard deviation of t-statistic
                                     # involving the prediction value
t0=qt(p=0.05/2, df=n-2, lower.tail = FALSE, log.p = FALSE)


LCB = y0 - (t0)*s0                   # Lower bound of the C.I
UCB = y0 + (t0)*s0                   # Upper bound of the C.I

cat("LCB = ", LCB, "      ", "UCB = ", UCB)


## LCB =   44.18       UCB =  74.43
```

Other R codes for reading a Prediction Interval

For a given x, the interval estimate of the dependent variable y is called the prediction interval. We can read this interval by using the function "predict":

```
LR = lm(y ~ x)                       # Linear regression model
data=data.frame(x,y)
newdata=data.frame(x=50)             # wrap the parameter

predict(LR,newdata,level = 0.95,interval="predict")


##     fit   lwr   upr
## 1 59.3 44.18 74.43
```

The 95% prediction interval of the Calculus Scores for the Motivational Test Score of 50 is between 44.17651 and 74.42924. The interval contains CS averages less than 50. This shows that although the MT may support students' preparation, it is not sufficient to ensure a high achievement in Calculus course. There might be other factors that affect students' success.

1.6.2 Sales Volume and Price

A company manager wants to establish a relationship between the sales of a certain product and the price. The company research department provides the following data:

Price in Dollars $= x$	35	40	45	48	50
Daily sales volume in units $= y$	80	75	68	66	63

i)   Find the least squares line of best fit of y as a function of x.

ii)   Plot the points and the regression line on the same graph.

iii)   Make a conjecture about the number of units that would be sold at a price of $60.

Solution. i) Using the "lm" function in R, we can obtain the equation of the best line fit.

```
x=c(35, 40, 45, 48, 50)          # price values
y=c(80, 75, 68, 66, 63)          # daily sales units
fit = lm(y ~ x)                  # Linear model
fit


##
## Call:
## lm(formula = y ~ x)
##
## Coefficients:
## (Intercept)             x
##      119.84         -1.13
```

The equation of the line is given by: $y = 119.845 - 1.134x$.

There is a decrease of almost 1.15 units of Daily sales volume for an increase of one dollar in the price.

ii) Figure 1.6 shows the plot of the data and the best line fit.

```
library(ggplot2)
mydata=data.frame(x,y)
ggplot(mydata,aes(x=x, y=y))+ geom_point() +
      geom_smooth(formula="y ~ x", method=lm, se=F) +
      xlab("Price in Dollars") + ylab("Daily sales volume in units")
```

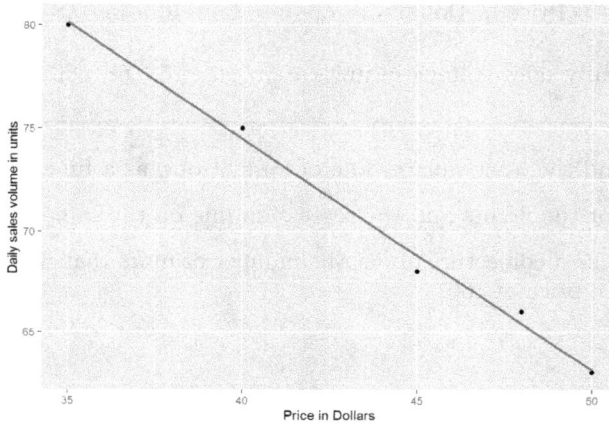

FIGURE 1.6 Daily sales versus Price

iii) Using the regression line, we can expect an approximate value of the number of units that would be sold at a price of $60. For this, we can use the function "predict" in R.

```
df=data.frame(x,y)
model= y ~ x                 # Store the model to estimate
fit = lm(model, data=df)     # Estimate the model with the data frame df
fit$coefficients
```

```
## (Intercept)          x
##     119.845      -1.134
```

```
newdata=data.frame(x=60)        # wrap the parameter
predict(fit,newdata)
```

```
##     1
## 51.8
```

```
# Confidence Interval
predict(fit,newdata,level = 0.95,interval="predict")
```

```
##    fit   lwr   upr
## 1 51.8 48.14 55.46
```

One can check these calculations using the detailed formulas:

```
x=c(35, 40, 45, 48, 50)
y=c(80, 75, 68, 66, 63)
n=length(x)
x0=60
y0= coefficients(fit)[1] + coefficients(fit)[2]*x0

sxy=sum((x-mean(x))*(y-mean(y)))
sxx=sum((x-mean(x))*(x-mean(x)))
syy=sum((y-mean(y))*(y-mean(y)))
SSR=sxy*sxy/sxx
SSE=syy-SSR
MSE=SSE/(n-2)

var=MSE*(1 + 1/n + (x0-mean(x))*(x0-mean(x))/sxx )

s0=sqrt(var)                    # standard deviation of the t-statistic

alpha = 0.05

# t-value
t0=qt(p=alpha/2, df=n-2, lower.tail = FALSE, log.p = FALSE)

LCB= y0 - (t0)*s0               #  Lower bound of the C.I
UCB= y0 + (t0)*s0               #  Upper bound of the C.I
LCB
```

```
## (Intercept)
##       48.14
```

UCB

```
## (Intercept)
##        55.46
```

A 95% confidence interval for the prediction of the Daily sales volume in units is $(48.14, 55.45)$ for the price of 60 dollars. The interval shows that the Daily sales will decrease quickly. Therefore, the company manager has to make a decision on the suitable price that maximizes the profit.

2

Multi-linear Regression

2.1 Description

Defining the Multiple Linear model

The Multiple linear regression model is an extension of simple linear regression. To explain the variation in a variable y, several independent variables (or predictors) are considered. The purpose is to make more accurate predictions.

Example. Systolic Blood Pressure y might be related to four factors; the person's:

x_1 : age, x_2 : weight, x_3 : height, x_4 : BMI (Body Mass Index).

A researcher would like to construct a prediction equation relating y, x_1, x_2, x_3, and x_4 by using the collected measurements.

Some questions arise:

- How well does the model fit?

- How strong is the relationship between y and the predictor variables x_i?

- Have any important assumptions been violated?

- How good are estimates and predictions?

DOI: 10.1201/9781003616832-2

The multi-linear model can be summarized as below (D.D. Wackerly, 2008):

We have a population of y_i values for each point $x_i = (x_{i1}, x_{i2}, \ldots, x_{ik})$,

    – the population random variable corresponding to x_i is $Y_i = Y_{x_i}$.

    – We assume we made n observations: $y_1, y_2, \ldots, y_n$.

We want to assume that

$$E(Y_i) = \beta_0 + \beta_1 x_{i1} + \beta_2 x_{i2} + \ldots + \beta_k x_{ik} \qquad \text{for each } x_i.$$

The errors of observations

$$\varepsilon_i = Y_i - (\beta_0 + \beta_1 x_{i1} + \beta_2 x_{i2} + \ldots + \beta_k x_{ik}) \qquad i = 1, 2, \ldots, n$$

represent the deviation between the value selected Y_i and the expected population mean.

We assume that $\varepsilon_1, \varepsilon_2, \ldots, \varepsilon_n$ are independent random variables with

$$E(\varepsilon_i) = 0 \qquad \text{and} \qquad Var(\varepsilon_i) = \sigma^2.$$

This multiple model has $(k + 1)$ unknown parameters: $\beta_0, \beta_1, \beta_2, \ldots \beta_k$, and σ.

β_i are called partial slopes or partial regression coefficients.

Finding the Minimizer

The least squares estimators of the unknown parameters $\beta_0, \beta_1, \beta_2, \ldots \beta_k$ are those values that minimize the multivariable function

$$Q(\beta_0, \beta_1, \beta_2, \ldots \beta_k) = \sum_{i=1}^{n} (y_i - (\beta_0 + \beta_1 x_{i1} + \beta_2 x_{i2} + \ldots + \beta_k x_{ik}))^2 = \sum_{i=1}^{n} \varepsilon_i^2$$

Since Q is a polynomial function of $\beta_0, \beta_1, \beta_2, \ldots \beta_k$, it is differentiable, and the minimizing values satisfy

$$\frac{\partial Q}{\partial \beta_0} = \frac{\partial Q}{\partial \beta_1} = \ldots = \frac{\partial Q}{\partial \beta_k} = 0 \qquad\qquad (*)$$

Fitting the Linear Model using Matrices

By introducing the following matrices

$$X = \begin{bmatrix} 1 & x_{11} & \cdots & x_{1k} \\ \vdots & \vdots & \cdots & \vdots \\ 1 & x_{n1} & \cdots & x_{nk} \end{bmatrix}$$

$$Y = \begin{bmatrix} y_1 \\ \vdots \\ y_n \end{bmatrix} \qquad \beta = \begin{bmatrix} \beta_0 \\ \vdots \\ \beta_k \end{bmatrix} \qquad \varepsilon = \begin{bmatrix} \varepsilon_1 \\ \vdots \\ \varepsilon_n \end{bmatrix},$$

we deduce the solution β of the necessary conditions $(*)$ as the solution of the matrix equation:

$$({}^t X.X)\widehat{\beta} = {}^t X.Y$$

where ${}^t A = (a_{ji})$ denotes the transpose matrix of the matrix $A = (a_{ij})$.

The least squares solution for the general linear model is therefore

$$\widehat{\beta} = ({}^t X.X)^{-1}.({}^t X.Y).$$

Simple Linear Regression Model: $y = \beta_0 + \beta_1 x_1 + \varepsilon$

$$ {}^t X.X = \begin{bmatrix} n & \sum x_i \\ \sum x_i & \sum x_i^2 \end{bmatrix}$$

$$ ({}^t X.X)^{-1} = \begin{bmatrix} \frac{\sum x_i^2}{n s_{xx}} & -\frac{\bar{x}}{s_{xx}} \\ -\frac{\bar{x}}{s_{xx}} & \frac{1}{s_{xx}} \end{bmatrix} = \begin{bmatrix} c_{00} & c_{01} \\ c_{01} & c_{11} \end{bmatrix}$$

$$ V(\widehat{\beta}_i] = c_{ii}\sigma^2 \qquad\qquad cov(\widehat{\beta}_0, \widehat{\beta}_1) = c_{01}\sigma^2$$

An unbiased estimator for σ^2 of the error ε is given by

$$S^2 = \frac{SSE}{n-2} \qquad \text{with}$$

$$SSE = \sum (y_i - \widehat{y}_i)^2 = {}^t YY - {}^t \widehat{\beta}({}^t XY) \qquad \text{and} \qquad \widehat{y}_i = \beta_0 + \beta_1 x_{i1}$$

Least-Squares Estimators

We recall the following **properties** (D.D. Wackerly, 2008):

- $E(\widehat{\beta}_i) = \beta_i, \qquad i = 0, 1, \ldots, k.$

- $V(\widehat{\beta}_i] = c_{ii}\sigma^2.$

- $cov(\widehat{\beta}_i, \widehat{\beta}_j) = c_{ij}\sigma^2,$

    c_{ij} is the element in row i and column j of the matrix $({}^tX.X)^{-1}$.

- An unbiased estimator for σ^2 is

$$S^2 = \frac{SSE}{n - (k + 1)} \qquad \text{with} \qquad SSE = {}^t YY - {}^t \widehat{\beta}({}^tXY),$$

    where SSE is the sum of squares of errors

$$SSE = \sum_{i=1}^{n}(y_i - \widehat{y}_i)^2 \qquad \text{and} \qquad \widehat{y}_i = \beta_0 + \beta_1 x_{i1} + \ldots + \beta_k x_{ik}.$$

If the **residuals** ε_i, for $i = 1, 2, \ldots, n$ are normally distributed, then

- $\widehat{\beta}_0, \widehat{\beta}_1, \ldots, \widehat{\beta}_k$ are normally distributed in repeated sampling.

- The random variable

$$\frac{[n - (k + 1)]S^2}{\sigma^2}$$

    has a χ^2 distribution with $(n - (k + 1))$ degrees of freedom.

- The statistics S^2 and $\widehat{\beta}_i$ are independent for each $i = 0, 1, \ldots, k.$

Solved Problems

2.1.1   Operations on Matrices with R

The following illustrates basic operations involving matrices.

Defining a matrix

```
A=matrix(c(1,2,3,4), nrow=2, ncol=2)
# nrow & ncol represent the number of rows and columns respectively.
A
```

```
##      [,1] [,2]
## [1,]    1    3
## [2,]    2    4
```

```
# reading the term in the 1st row, 2nd column of A
a12=A[1,2]
a12
```

```
## [1] 3
```

Transpose of a matrix $\mathbf{A} = (a_{ij})$ is the matrix ${}^t A = (a_{ji})$

```
t(A)                                    # gives the transpose of A
```

```
##      [,1] [,2]
## [1,]    1    2
## [2,]    3    4
```

Product ^{t}AA

```
(t(A))%*%A          # the operator product between two matrices is: %*%
```

```
##      [,1] [,2]
## [1,]    5   11
## [2,]   11   25
```

Solve a linear system: Ax = b

```
b=c(5,6)
x=solve(A,b)       #gives the solution x of the linear system Ax=b
x
```

```
## [1] -1   2
```

Transform a data.frame to a matrix

```
x1=c(1,2)
x2=c(3,4)
DF=data.frame(x1,x2)
M=data.matrix(DF, rownames.force = NA)
M
```

```
##      x1 x2
## [1,]  1  3
## [2,]  2  4
```

2.1.2 Best Line Fit Through 5 Points

1. Use the method of least squares to fit a straight line to the $n = 5$ data points:

$$(x, y): \quad (-2, 0), \quad (-1, 0), \quad (0, 1), \quad (1, 1), \quad (2, 3)$$

2. Find the variances of the estimators $\widehat{\beta}_0$, $\widehat{\beta}_1$, and $cov(\widehat{\beta}_0, \widehat{\beta}_1)$.

3. Find an estimator of σ^2.

Solution.   1. To find $\widehat{\beta}$, we solve the equation: $({}^t X.X)\widehat{\beta} = {}^t X.Y$

```
X=matrix(c(1,1,1,1,1,-2,-1,0,1,2), nrow=5, ncol=2)
Y=c(0,0,1,1,3)
beta=solve(t(X)%*%X, t(X)%*%Y)        # partial slopes
beta
```

```
##         [,1]
## [1,]   1.0
## [2,]   0.7
```

The regression line is: $y = 1.0 + 0.7x$. Using the "lm" function, we obtain the same line equation. Figure 2.1 illustrates the approximation.

```
x=c(-2,-1,0,1,2)          # 1st coordinates of the points
y=c(0,0,1,1,3)            # 2nd coordinates of the points
fit = lm(y ~ x)
plot(x,y, col="blue", main="Fitting the data to a Line",lwd=2,
     col.main="darkgreen")
curve(coefficients(fit)[1] + coefficients(fit)[2]*x,-3,3,
        add=TRUE, col="red", lwd=2)
legend('topleft',inset=0.05,c("y = 1 + 0.7x"),
        lty=1,col=c("red"),title="best line fit")
```

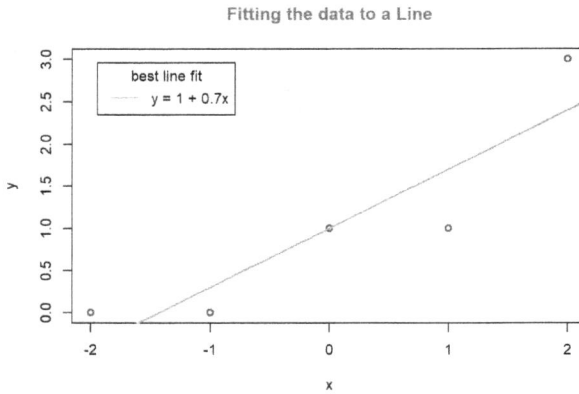

FIGURE 2.1 Best line fit

2. The variances of the estimators $\widehat{\beta}_0$, $\widehat{\beta}_1$:

```
A=t(X)%*%X
I=matrix(c(1,0,0,1), nrow=2, ncol=2)        # The 2x2 identity matrix
C=solve(A,I)                    # C is the solution of A C = C A = I
C
```

```
##        [,1] [,2]
## [1,]   0.2  0.0
## [2,]   0.0  0.1
```

Thus
$$cov(\widehat{\beta}_0, \widehat{\beta}_1) = c_{01}\sigma^2 = 0$$

$$V(\widehat{\beta}_0) = c_{00}\sigma^2 = 0.2\sigma^2 \qquad\qquad V(\widehat{\beta}_1) = c_{11}\sigma^2 = 0.1\sigma^2$$

3. An estimator of σ^2 is given by $S^2 = SSE/(n - (k+1))$

```
SSE = t(Y)%*%Y - (t(beta))%*%(t(X)%*%Y)
n=5
k=1
S2= SSE/(n-(k+1))              # estimator of population variance
S2
```

```
##        [,1]
## [1,] 0.3667
```

2.1.3 Fit a Parabola to the Data
$$(x, y): \quad (-2, 0), \quad (-1, 0), \quad (0, 1), \quad (1, 1), \quad (2, 3)$$

Solution. We are looking for the Model: $y = \beta_0 + \beta_1 x + \beta_2 x^2 + \varepsilon$

```
X=matrix(c(1,1,1,1,1, -2,-1,0,1,2,4,1,0,1,4), nrow=5, ncol=3)
Y=c(0,0,1,1,3)                              # y-coordinates
beta=solve(t(X)%*%X, t(X)%*%Y)              # partial slopes
beta
```

```
##         [,1]
## [1,] 0.5714
## [2,] 0.7000
## [3,] 0.2143
```

The model is: $y = 0.5714286 + 0.7x + 0.2142857x^2$.

Other R codes. We apply the "lm" function to y as a function of x and x^2, and save the linear regression model in a new variable "fit". Then we read the parameters of the estimated regression equation by calling "fit".

```
x = c(-2, -1, 0, 1, 2)                      # x-coordinates
x2 = x*x
DF = data.frame(x,x2,Y)
fit = lm(Y ~ x + x2, data = DF)
fit
```

```
##
## Call:
## lm(formula = Y ~ x + x2, data = DF)
##
## Coefficients:
## (Intercept)            x            x2
##       0.571        0.700         0.214
```

Figure 2.2 illustrates the parabolic approximation.

```
plot(x,Y, col="blue", pch = 15,main="Fitting the data to a
     parabola",lwd=2, col.main="darkgreen")
curve( 0.5714 + 0.7*x + 0.2143*x^2, -3,3, add=TRUE, col="red", lwd=2)
legend('topleft',inset=0.05, c("y = 0.5714 + 0.7x + 0.2143 x^2"),
       lty=1,col=c("red"),title="best parabola fit")
```

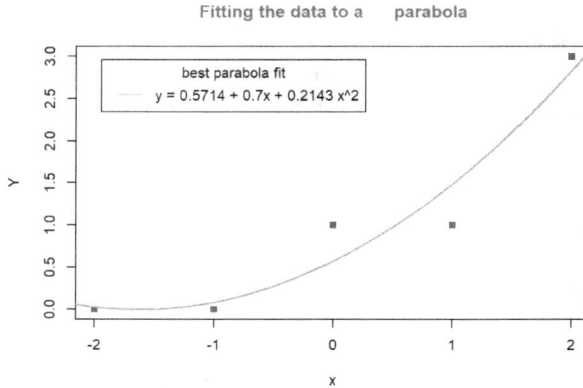

FIGURE 2.2 Parabolic Model-5 points

The sum of squares of errors is given by:

```
SSE = t(Y)%*%Y - ( t(beta))%*%(t(X)%*%Y )
n=5
k=2
S2= SSE/(n-(k+1))
S2
```

```
##           [,1]
## [1,] 0.2286
```

Note that

$$SSE_{Line-Model} = 0.3666667 > 0.2285714 = SSE_{Parabolic-Model}.$$

The Parabolic model is a better fit to the five points than the Line model since the Parabolic model has a smaller error.

TABLE 2.1 Lexus automobile Sales

Year, x	sales, y
1996	18.5
1997	22.6
1998	27.2
1999	31.2
2000	33.0
2001	44.9
2002	49.4
2003	35.0

2.1.4 Lexus Automobiles Sales

The manufacturer of Lexus automobiles has steadily increased sales since the 1989 launch of that brand in the United States. However, the rate of increase changed in 1996 when Lexus introduced a line of trucks. The sales of Lexus vehicles from 1996 to 2003 are shown in the accompanying Table 2.1.

a. Letting Y denote sales and x denote the coded year (-7 for 1996, -5 for 1997, through 7 for 2003), fit the model $Y = \beta_0 + \beta_1 x + \varepsilon$.

b. For the same data, fit the model $Y = \beta_0 + \beta_1 x + \beta_2 x^2 + \varepsilon$.

(This problem is formulated from D.D. Wackerly (2008) Exercise 11.69.)

Solution. a. We apply the "lm" function to y as a function of x and save the linear regression model in a new variable "fit1". Then we read the parameters of the estimated regression equation by calling "fit1".

```
x=c(-7, -5, -3, -1, 1, 3, 5, 7)                 # coded years
y=c(18.5, 22.6, 27.2, 31.2, 33, 44.9, 49.4, 35) # Lexus sales

data1=data.frame(x,y)
```

```
fit1 = lm(y ~ x, data=data1)
fit1
```

```
##
## Call:
## lm(formula = y ~ x, data = data1)
##
## Coefficients:
## (Intercept)                    x
##        32.72                 1.81
```

The linear model shows that there is an increase of almost 2% each year in the sales of Lexus vehicles from 1996 to 2003.

Figure 2.3 illustrates the linear model.

```
plot(x,y,   col = "blue", main="Fitting the data to a Line",
     col.main="darkgreen",lwd = 2, xlab="x: year", ylab="y: sales")
curve(coefficients(fit1)[1] + coefficients(fit1)[2]*x, -20,20,
                           add=TRUE, col="red", lwd=2)
legend('topleft',inset=0.05, c("y = 32.725 + 1.812 x"),
                    lty=1,col=c("red"),title="best line fit")
```

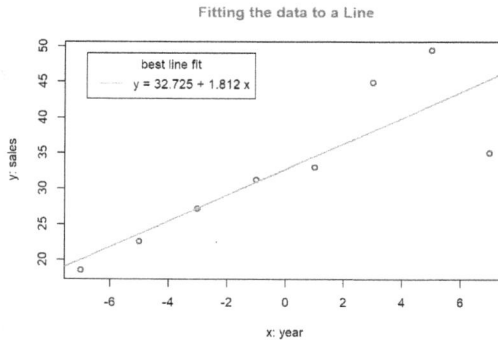

FIGURE 2.3 Lexus sales-Linear Model

b. We apply the "lm" function to y as a function of x and x^2, and save the linear regression model in a new variable "fit2". Then we read the parameters of the estimated regression equation by calling "fit2".

```
x2=x*x
data2=data.frame(x,x2,y)
fit2 = lm(y ~ x + x2, data=data2)
fit2


##
## Call:
## lm(formula = y ~ x + x2, data = data2)
##
## Coefficients:
## (Intercept)              x             x2
##       35.562          1.812         -0.135
```

Figure 2.4 illustrates the parabolic approximation.

```
plot(x,y, col="blue", main="Fitting the data to a parabola",
     col.main="darkgreen",lwd=2, xlab="x: year", ylab="y: sales")
curve(35.5625+1.8119*x-0.1351*x^2,from=-8,to=8,add=TRUE,col="red",lwd=2)
legend('topleft',inset=0.05, c("y = 35.5625 + 1.8119 x - 0.1351 x^2"),
                 lty=1,col=c("red"),title="best parabolic fit")
```

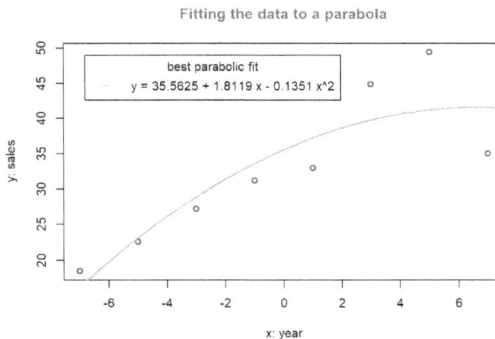

FIGURE 2.4 Lexus sales-Parabolic Model

2.2   Analysis of Variance for a Multi-linear Regression

ANOVA table: **AN**alyzis **O**f **VA**riance for multilinear regression

Source of variation in y	df: degrees of freedom	SS: Sum of Squares	MS = SS/df Mean of SS	F
Regression explained variation	k	$SSR = SST - SSE$	$MSR = \dfrac{SSR}{k}$	$F = \dfrac{MSR}{MSE}$
Error unexplained variation	$n-(k+1)$	$SSE = {}^{t}YY - {}^{t}\widehat{\beta}({}^{t}XY)$	$MSE = \dfrac{SSE}{n-(k+1)} = s^2$	
Total variation in y	$n-1$	$SST = s_{yy}$		

— The total variation in the response variable y

$$\text{Total}SS \;=\; SST \;=\; s_{yy} \;=\; \sum_{i=1}^{n}(y_i - \overline{y})^2, \qquad\qquad \overline{y} = \frac{1}{n}\sum_{i=1}^{n}y_i$$

— The sum of squares of regression **SSR** measures the amount of the variation explained by using the multi-linear regression

$$SSR \;=\; \sum_{i=1}^{n}(\widehat{y}_i - \overline{y})^2 = \sum_{i=1}^{n}(\beta_0 + \beta_1 x_{i1} + \ldots + \beta_k x_{ik} - \overline{y})^2$$

— The sum of squares of errors **SSE** measures the "residual variation" in the data that is not explained by the independent variables $x_1, x_2, \ldots, x_n$

$$SSE \;=\; SST \;-\; SSR$$

− The **mean squares for errors**

$$MSE \;=\; s^2 \;=\; \frac{SSE}{n-(k+1)}$$

The Coefficient of Determination R^2

The ratio

$$R^2 = \frac{SSR}{SST} = \frac{\text{Explained SS}}{\text{Total SS}}; \qquad 0 \leqslant R^2 \leqslant 1,$$

serves as a measure of "goodness of fit"; it represents the proportion of variation in y that can be explained by the linear influence of the variables $x_1, x_2, \ldots, x_k$. The value of R^2 gets closer to 1 as the relationship gets stronger.

The Analysis of Variance F-test

To test the contribution of the predictor variables $x_1, x_2, \ldots, x_k$ in the regression model better than the simple predictor $\bar{y}$, we perform the F test with the hypotheses:

$$H_0: \quad \beta_1 = \beta_2 = \ldots = \beta_k = 0 \qquad\qquad \text{versus}$$

$H_a:$ at least one of the parameters $\beta_1, \beta_2, \ldots, \beta_k$ differs from 0.

The test statistic is found in the ANOVA table as

$$F = \frac{MSR}{MSE} \qquad \rightsquigarrow \qquad \mathcal{F}_{df_1=k \;;\; df_2=n-(k+1)} \qquad \text{distributed}$$

Rejection region. *With a Level of significance α, the null Hypothesis H_0 is rejected if $F > F_\alpha$.*

$$\alpha = P\Big(H_0 \text{ is rejected } \Big| H_0 \text{ is true }\Big) = P\Big(F > F_\alpha\Big)$$

TABLE 2.2 Rubber's resistance

Y	x1	x2
83	1.0	-1.0
113	1.0	1.0
92	-1.0	1.0
82	-1.0	-1.0
100	0.0	0.0
96	0.0	0.0
98	0.0	0.0
95	0.0	1.5
80	0.0	-1.5
100	1.5	0.0
92	-1.5	0.0

Solved Problems

2.2.1 Rubber's Resistance

It is desired to relate abrasion resistance of rubber Y to the amount of silica filler x_1 and the amount of coupling agent x_2. Fine particle silica fibers are added to rubber to increase strength and resistance to abrasion. The coupling agent chemically bonds the filler to the rubber polymer chains and thus increases the efficiency of the filler (see Table 2.2).

1. Fit the data to the second order model

$$y = \beta_0 + \beta_1 x_1 + \beta_2 x_2 + \beta_3 x_1^2 + \beta_4 x_2^2 + \beta_5 x_1 x_2 + \varepsilon.$$

2. Give the ANOVA table for the multiple regression of this model.

3. Evaluate the usefulness of this model.

(This problem is formulated from D.D. Wackerly (2008) Exercise 11.19.)

Solution.

1. Fit the second-order model to the data. We use matrix equations to fit the model.

```
u=c(1, 1, 1, 1, 1, 1, 1, 1, 1, 1, 1)
x1=c(1,1,-1,-1,0,0,0,0,0,1.5,-1.5)        # amount of silica filler
x2=c(-1,1,1,-1,0,0,0,1.5,-1.5,0,0)        # amount of coupling agent
x1x1 = x1*x1
x2x2 = x2*x2
x1x2 = x1*x2


DF=data.frame(u,x1,x2, x1x1, x2x2, x1x2)

X=data.matrix(DF, rownames.force = NA)
X
```

```
##        u    x1    x2 x1x1 x2x2 x1x2
##  [1,] 1  1.0 -1.0 1.00 1.00   -1
##  [2,] 1  1.0  1.0 1.00 1.00    1
##  [3,] 1 -1.0  1.0 1.00 1.00   -1
##  [4,] 1 -1.0 -1.0 1.00 1.00    1
##  [5,] 1  0.0  0.0 0.00 0.00    0
##  [6,] 1  0.0  0.0 0.00 0.00    0
##  [7,] 1  0.0  0.0 0.00 0.00    0
##  [8,] 1  0.0  1.5 0.00 2.25    0
##  [9,] 1  0.0 -1.5 0.00 2.25    0
## [10,] 1  1.5  0.0 2.25 0.00    0
## [11,] 1 -1.5  0.0 2.25 0.00    0
```

```
# Abrasion resistance of rubber
Y=c(83, 113, 92, 82, 100, 96, 98, 95, 80, 100, 92)

# Partial regression coefficients
beta=solve(t(X)%*%X, t(X)%*%Y)
beta
```

```
##            [,1]
## u       98.0046
## x1       4.0000
## x2       7.3529
## x1x1    -0.8788
## x2x2    -4.6565
## x1x2     5.0000
```

The second-order model is given by:

$$Y = 98.0045558 + 4x_1 + 7.3529412x_2 - 0.8787649x_1^2 - 4.6565426x_2^2 + 5x_1x_2$$

Other R codes

• We apply the "lm" function to express y as a function of x_1, x_2, x_1^2, x_2^2, x_1x_2, and save the linear regression model in a new variable "fit". Then, we read the parameters of the estimated regression equation by calling "fit".

```
fit = lm(Y ~ x1 + x2 + x1x1 + x2x2 + x1x2, data=DF)
fit
```

```
##
## Call:
## lm(formula = Y ~ x1 + x2 + x1x1 + x2x2 + x1x2, data = DF)
##
## Coefficients:
## (Intercept)           x1           x2         x1x1
##      98.005        4.000        7.353       -0.879
##        x2x2         x1x2
##      -4.657        5.000
```

• We apply the function "summary" to read more information on the regression.

```
summary(fit)
```

```
##
## Call:
## lm(formula = Y ~ x1 + x2 + x1x1 + x2x2 + x1x2, data = DF)
##
## Residuals:
##        1        2        3        4        5        6
## -1.11631  4.17781  1.17781 -4.11631  1.99544 -2.00456
##        7        8        9       10       11
## -0.00456 -3.55675  3.50208 -2.02733  1.97267
##
## Coefficients:
##               Estimate Std. Error t value Pr(>|t|)
## (Intercept)     98.005      2.269   43.19 1.3e-07 ***
## x1               4.000      1.354    2.95  0.0318 *
## x2               7.353      1.354    5.43  0.0029 **
## x1x1            -0.879      1.527   -0.58  0.5899
```

```
## x2x2              -4.657      1.527    -3.05    0.0285 *
## x1x2               5.000      1.974     2.53    0.0524 .
## ---
## Signif. codes:
## 0 '***' 0.001 '**' 0.01 '*' 0.05 '.' 0.1 ' ' 1
##
## Residual standard error: 3.95 on 5 degrees of freedom
## Multiple R-squared:  0.915,  Adjusted R-squared:  0.831
## F-statistic: 10.8 on 5 and 5 DF,  p-value: 0.0103
```

2. ANOVA table for the multiple regression of the model.

The sum of squares of errors is given by:

```
SSE = t(Y)%*%Y - (t(beta))%*%(t(X)%*%Y )
SSE
```

```
##          [,1]
## [1,] 77.95
```

The ANOVA Table:

```
k=5
n=11

TotalSS = sum(Y*Y) - (sum(Y))^2/n
SSR = TotalSS - SSE
MSR = SSR/(n-(k+1))
MSE = SSE/(n-(k+1))


V1=c("Regression", "Error")
V2=c(k,n-(k+1))
V3=c(SSR, SSE)
V4=c(MSR,MSE)
F=c(MSR/MSE, ". " )
ANOVA=data.frame(Source=V1, df=V2,  SS=V3,  MS=V4, Fvalue=F )
ANOVA
```

```
##        Source df    SS     MS            Fvalue
## 1 Regression  5 844.23 168.85 10.8307294311932
## 2      Error  5  77.95  15.59                 .
```

3.i) Testing the usefulness of the Regression Model through the **coefficient of determination** R^2.

How well does the regression model fit?

The ANOVA table provides a statistical measure of the strength of the model in the coefficient of determination: R^2—the proportion of the total variation that is explained by the regression of Y on the predictor variables x_1, x_2, x_1^2, x_2^2, x_1x_2—defined as

$$R^2 = \frac{SSR}{TotalSS} = \frac{844.23381}{77.94801 + 844.23381} = 0.9154744 \qquad \text{or} \qquad 91.54\%$$

```
SST=SSR+SSE        # TotalSS
R2=SSR/SST
R2
```

```
##           [,1]
## [1,] 0.9155
```

Hence, for this example, 91.54% of the total variation has been explained by the regression model. The model fits very well.

3.ii) Testing the usefulness of the Regression Model through the **Analysis of Variance F-test.**

Is the regression equation that uses information provided by the predictor variables x_1, x_2, x_1^2, x_2^2, x_1x_2 substantially better than the simple predictor $\overline{Y}$ that does not rely on any of the x values?

This question is answered using an overall F-test with the hypotheses

$$H_0: \quad \beta_1 = \beta_2 = \ldots = \beta_5 = 0 \qquad \text{versus}$$

$$H_a: \quad \text{at least one of the parameters } \beta_1, \beta_2, \ldots, \beta_5 \text{ differs from 0.}$$

The test statistic is found in the ANOVA table as

```
F=MSR/MSE
F
```

```
##         [,1]
## [1,] 10.83
```

which has an F distribution with $df_1 = k = 5$ and $df_2 = n-k-1 = 11-5-1 = 5$.

The p-value is given in the summary and can be calculated by:

```
k=5
n=11
pv=pf(MSR/MSE, df1= k, df2=n-k-1, lower.tail=FALSE)
pv
```

```
##         [,1]
## [1,] 0.01028
```

Since the p-value is less than 0.05, we can declare the regression to be highly significant. That is, at least one of the predictor variables is contributing significant information for the prediction of the response variable Y.

Other R codes

```
anova(fit)
```

```
## Analysis of Variance Table
##
## Response: Y
##
            Df Sum Sq Mean Sq F value  Pr(>F)
## x1         1    136     136    8.72  0.0318 *
## x2         1    460     460   29.48  0.0029 **
## x1x1       1      4       4    0.24  0.6449
## x2x2       1    145     145    9.30  0.0285 *
## x1x2       1    100     100    6.41  0.0524 .
## Residuals  5     78      16
## ---
## Signif. codes:
## 0 '***' 0.001 '**' 0.01 '*' 0.05 '.' 0.1 ' ' 1
```

```
newDF = data.frame(x1, x2, x1x1, x2x2, x1x2, Y)
Model0 = lm(Y ~ 1, newDF)
Model1 = lm(Y ~ x1 + x2 + x1x1 + x2x2 + x1x2, newDF)
anova(Model0, Model1)
```

```
## Analysis of Variance Table
##
## Model 1: Y ~ 1
## Model 2: Y ~ x1 + x2 + x1x1 + x2x2 + x1x2
##   Res.Df RSS Df Sum of Sq    F Pr(>F)
## 1      10 922
## 2       5  78  5       844 10.8   0.01 *
## ---
## Signif. codes:
## 0 '***' 0.001 '**' 0.01 '*' 0.05 '.' 0.1 ' ' 1
```

2.3 Inferences on a Linear Combination of the Parameters

We wish to make an inference about the linear combination of the β_i:

$$a_0\beta_0 + a_1\beta_1 + \ldots + a_k\beta_k = {}^t a.\beta, \qquad \text{where} \qquad a = \begin{bmatrix} a_0 \\ \vdots \\ a_k \end{bmatrix}$$

and $a_0, a_1, \ldots, a_k$ are constants.

Properties.

– An unbiased estimator for ${}^t a.\beta$ is ${}^t a.\widehat{\beta}$ and we have:

$$E({}^t a.\widehat{\beta}) = {}^t a.\beta \qquad \text{and} \qquad V({}^t a\widehat{\beta}) = [{}^t a({}^t X.X)^{-1} a]\,\sigma^2.$$

– ${}^t a.\beta$ is normally distributed in repeated sampling,

$$Z = \frac{{}^t a\widehat{\beta} - ({}^t a\beta)_0}{\sigma\sqrt{{}^t a({}^t XX)^{-1} a}} \quad \rightsquigarrow \quad \mathcal{N}(0,1)$$

– If we substitute S with σ, then

$$T = \frac{{}^t a\widehat{\beta} - {}^t a\beta}{S\sqrt{{}^t a({}^t XX)^{-1} a}} \quad \rightsquigarrow \quad \mathcal{T}_{n-(k+1)}$$

Testing ^{t}a.β. We formulate here the case involving a t-student distribution. The case of a normal distribution can be deduced similarly.

– Level of Significance:

$$\alpha = P\Big(H_0 \text{ is rejected } \Big| H_0 \text{ is true}\Big) = P\Big(y \in C \Big| H_0 \text{ is true}\Big)$$

– Test Statistic: For $x_1, x_2, \ldots, x_n$ fixed and for an observed sample $y_1, y_2, \ldots, y_n$, the observed statistic is given by:

$$t \;=\; \frac{{}^t a \widehat{\beta} - ({}^t a \beta)_0}{s \sqrt{{}^t a ({}^t X X)^{-1} a}}$$

– Tests: The following are the possible statistical tests that we consider.

Null Hypothesis	Alternative Hypothesis	Test	Rejection Region C
H_0: ${}^t a \beta \leqslant ({}^t a \beta)_0$ or ${}^t a \beta = ({}^t a \beta)_0$	H_a : ${}^t a \beta > ({}^t a \beta)_0$	$P(T > t_{\alpha, n-(k+1)}) = \alpha$	$\{y : \; t > t_{\alpha, n-(k+1)}\}$
H_0: ${}^t a \beta \geqslant ({}^t a \beta)_0$ or ${}^t a \beta = ({}^t a \beta)_0$	H_a : ${}^t a \beta < ({}^t a \beta)_0$	$P(T < -t_{\alpha, n-(k+1)}) = \alpha$	$\{y : \; t < -t_{\alpha, n-(k+1)}\}$
H_0 :${}^t a \beta = ({}^t a \beta)_0$	H_a : ${}^t a \beta \neq ({}^t a \beta)_0$	$P(|T| > t_{\frac{\alpha}{2}, n-(k+1)}) = \alpha$	$\{y : \; |t| > t_{\frac{\alpha}{2}, n-(k+1)}\}$

The three tests are respectively described as "right", "left", and "two" sided tests.

– Decision Rule

Reject H_0 if the observed sample falls in the rejection region C. With the decision rule, we achieve a level of significance of α; that is, $\alpha\%$ is the mistake made of rejecting H_0 when H_0 is actually true.

Confidence Interval

The random C.I is characterized by

$$P\left(\left|\frac{{}^t a\widehat{\beta} - {}^t a\beta}{S\sqrt{{}^t a({}^t XX)^{-1}a}}\right| < t_{\alpha/2,n-(k+1)}\right) = 1 - \alpha.$$

For predicting a particular value of ${}^t a\beta$, we use an $(1-\alpha)100\%$ confidence and prediction interval

$$
\begin{aligned}
{}^t a\widehat{\beta} - t_{\alpha/2,n-(k+1)} S\sqrt{{}^t a({}^t XX)^{-1}a} &< {}^t a\beta \\
&< {}^t a\widehat{\beta} + t_{\alpha/2,n-(k+1)} S\sqrt{{}^t a({}^t XX)^{-1}a}.
\end{aligned}
$$

Solved Problems

2.3.1   Reliability of a Parabolic Model: $y = \beta_0 + \beta_1 x + \beta_2 x^2 + \varepsilon$

1. Fit a parabola to the data:

$$(x, y) : \ (-3, 1), \ (-2, 0), \ (-1, 0), \ (0, -1), \ (1, -1), \ (2, 0), \ (3, 0)$$

2. Plot the points and sketch the fitted parabola as a check on the calculations.

3. Do the data present sufficient evidence to indicate curvature in the response function? Test using $\alpha = 0.05$ and give bounds to the attained significance level.

Solution.   1. Finding the Parabolic Model.

```
X=matrix(c(1,1,1,1,1,1,1, -3,-2,-1,0,1,2,3, 9,4,1,0,1,4,9),
          nrow=7, ncol=3)
Y=c(1, 0, 0, -1, -1, 0, 0)        # y-coordinates
beta=solve(t(X)%*%X, t(X)%*%Y)
beta                              # partial slopes
```

```
##             [,1]
## [1,] -0.7143
## [2,] -0.1429
## [3,]  0.1429
```

The model is: $y = -0.7142857 - 0.1428571x + 0.1428571x^2$.

2. Figure 2.5 illustrates the points and the parabola.

```
Xp=c(-3, -2, -1, 0, 1, 2, 3)
Y =c(1, 0, 0, -1, -1, 0, 0)
plot(Xp,Y,  col = "blue", pch=15, lwd=2, main="Parabolic Model",
      col.main="darkgreen")
curve(-0.7143-0.1429*x+0.1429*x^2,-3.5,3.5,add=TRUE,col="red",lwd=2)
legend('topright',inset=0.05,c("y=-0.7143-0.1429*x+0.1429*x^2"),
            lty=1,col=c("red"),title="best curve fit")
```

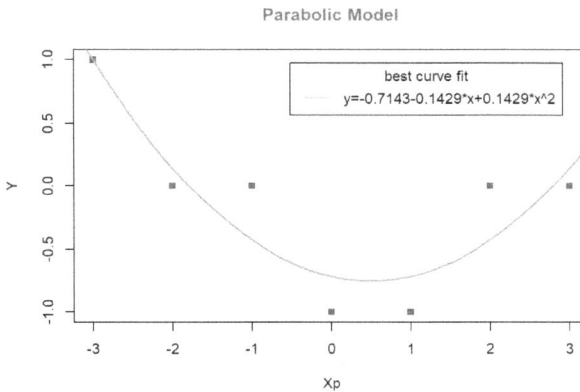

FIGURE 2.5 Parabolic Model-7 points

3. Test of the Validity of the Parabolic Model. We would like to check whether the parabolic model is a realistic description of the data. We test the hypothesis:

$$H_0 : \beta_2 = 0 \qquad \text{versus} \qquad H_a : \beta_2 \neq 0$$

Using matrix notations, we can write:

$$a = \begin{bmatrix} 0 \\ 0 \\ 1 \end{bmatrix} \qquad \beta = \begin{bmatrix} \beta_0 \\ \beta_1 \\ \beta_2 \end{bmatrix} \qquad \beta_2 = {}^t a.\beta$$

```
a=c(0,0,1)
beta2=t(a)%*%beta
beta2
```

```
##          [,1]
## [1,] 0.1429
```

The statistic $\quad T = \dfrac{{}^t a\widehat{\beta} - {}^t a\widehat{\beta_0}}{S\sqrt{{}^t a({}^t XX)^{-1} a}} \quad\quad \rightsquigarrow \quad \mathcal{T}_{n-(k+1)}$

is a Student t-distribution in repeated sampling with $df = n - (k+1)$ where

$$
{}^t a({}^t X.X)^{-1} a = {}^t a \begin{bmatrix} c_{00} & c_{01} & c_{02} \\ c_{10} & c_{11} & c_{12} \\ c_{20} & c_{21} & c_{22} \end{bmatrix} a = c_{22},
$$

$$
V({}^t a\widehat{\beta}) = {}^t a({}^t X.X)^{-1} a \sigma^2 = c_{22}\sigma^2
$$

Calculating c_{22}

```
A=t(X)%*%X
I=matrix(c(1,0,0,0,1,0,0,0,1), nrow=3, ncol=3)
C=solve(A,I)                                    # C is the inverse of A
C
```

```
##            [,1]     [,2]      [,3]
## [1,]   0.33333 0.00000 -0.04762
## [2,]   0.00000 0.03571  0.00000
## [3,]  -0.04762 0.00000  0.01190
```

```
c22=C[3,3]
c22
```

```
## [1] 0.0119
```

Calculating S   An estimator of σ^2 is

$$
S^2 = \frac{SSE}{n-(k+1)} \qquad \text{with} \qquad SSE = \sum(y_i - \hat{y}_i)^2 = {}^t YY - {}^t \widehat{\beta}({}^t XY)
$$

```
SSE = t(Y)%*%Y - (t(beta))%*%(t(X)%*%Y )
k=2
n=7
S2 = SSE/(n-(k+1))
S = sqrt(S2)
cat("The value of S is ", S)
```

```
## The value of S is  0.378
```

The observed statistic   is given by $t = \dfrac{\widehat{\beta}_2 - \beta_2}{S\sqrt{c_{22}}}$

```
t=(beta2-0)/(S*sqrt(c22))
t
```

```
##         [,1]
## [1,] 3.464
```

▶ **The Rejection Region Approach.** With the level of significance $\alpha = 0.05$, the t-value:

$$t_{\alpha/2;df=n-(k+1)} = t_{0.025;7-3=4} = t_0$$

```
t0=qt(p=0.05/2, df=n-(k+1),lower.tail = FALSE, log.p = FALSE)
t0
```

```
## [1] 2.776
```

and the rejection region is characterized by $|t| > t_{\alpha/2} = 2.776445$.

The observed value t satisfies $|t| = 3.464102 > t_{\alpha/2} = 2.776445$; that is, t falls in the rejection region. We can reject H_0. The data presents sufficient evidence to indicate that the model is nonlinear.

▶ **The p-value Approach.** Because the test is two tailed, the p-value equals

$$p\text{-value} = 2 \times P(|T| > t)$$

```
pv=2*pt(q=abs(t), df=n-(k+1), lower.tail = FALSE, log.p = FALSE)
pv
```

```
##         [,1]
## [1,] 0.02572
```

The p-value is less than the level of significance α. H_0 is rejected.

▶ The Confidence Interval Approach.

The C.I is characterized by: $1 - \alpha = P(|T| < t_{\alpha/2})$

$$= P\left({}^t a\widehat{\beta} - t_{\alpha/2}S\sqrt{{}^t a({}^tX.X)^{-1}a} <^t a\beta <^t a\widehat{\beta} + t_{\alpha/2}S\sqrt{{}^t a({}^tX.X)^{-1}a}\right).$$

A $100(1 - \alpha)\%$ *confidence interval for* ${}^t a\beta$ *is*

$$ {}^t a\widehat{\beta} \pm t_{\alpha/2}S\sqrt{{}^t a({}^tX.X)^{-1}a} $$

```
LCB= beta2 - t0*S*sqrt(c22)        # Lower Confidence Bound
UCB= beta2 + t0*S*sqrt(c22)        # Upper Confidence Bound
cat("LCB =", LCB, "        ;    ", "UCB=", UCB)
```

```
## LCB = 0.02836       ;     UCB= 0.2574
```

The calculated interval is $(0.02835848, 0.2573558)$. *The interval doesn't include the value 0; the value hypothesized in* H_0. *This indicates that it is quite possible that* $\beta_2 \neq 0$, *and therefore agrees with the test conclusions of the critical and p-value approaches: Reject* $H_0 : \beta_2 = 0$.

We notice the agreement between the conclusions reached by the three approaches.

■ Other R codes

We apply the "lm" function to y as a function of x and x^2, and save the linear regression model in a new variable "fit". Then we read the parameters of the estimated regression equation by calling "fit".

```
x=c(-3, -2, -1, 0, 1, 2, 3)
xx=x*x
y =c(1, 0, 0, -1, -1, 0, 0)
DF=data.frame(x,xx,y)
fit = lm(y ~ x + xx, data=DF)
```

```
summary(fit)
```

```
##
## Call:
## lm(formula = y ~ x + xx, data = DF)
##
```

```
## Residuals:
##         1         2         3         4         5
##   1.32e-16 -1.43e-01  4.29e-01 -2.86e-01 -2.86e-01
##         6         7
##   4.29e-01 -1.43e-01
##
## Coefficients:
##               Estimate Std. Error t value Pr(>|t|)
## (Intercept)   -0.7143     0.2182   -3.27    0.031 *
## x             -0.1429     0.0714   -2.00    0.116
## xx             0.1429     0.0412    3.46    0.026 *
## ---
## Signif. codes:
## 0 '***' 0.001 '**' 0.01 '*' 0.05 '.' 0.1 ' ' 1
##
## Residual standard error: 0.378 on 4 degrees of freedom
## Multiple R-squared:   0.8,   Adjusted R-squared:   0.7
## F-statistic:    8 on 2 and 4 DF,  p-value: 0.04
```

Since the p-value in the test $\beta_2 = 0$ is 0.0257 and is less than 0.05, then the variable x^2 is statistically significant in the multiple linear regression model. It is possible to have $\beta_2 \neq 0$.

TABLE 2.3 CO_2 measurements

Month	CO_2	Month	CO_2
1	386.9	7	387.7
2	387.4	8	385.9
3	388.8	9	384.8
4	389.5	10	384.4
5	390.2	11	386.0
6	389.4	12	387.3

2.3.2   Reliability of a Sinusoidal Model

Measurements of CO_2 in the atmosphere have been taken regularly over the last 50 years at the Mauna Loa Observatory in Hawaii. In addition to a general upward trend, the CO_2 data also has an annual cyclic behavior. Table 2.3) has the monthly measurements (in parts per million) for 2009.

1. Fit the model $y = \beta_0 + \beta_1 x + \beta_2 \sin(x\pi/6) + \varepsilon$ to the data, where x is time in months and y is the corresponding measurement of CO_2.

2. Plot the points and sketch the fitted curve as a check on the calculations.

3. Do the data present sufficient evidence to indicate curvature in the response function?

Test using $\alpha = 0.05$ and give bounds to the attained significance level.

(This problem is formulated from Holt (2013) Question 56, Section 8.5.)

Solution.

1. Finding the Sinusoidal Model.

```
u=c(1, 1 , 1, 1, 1, 1, 1, 1, 1, 1, 1, 1)
x=c(1, 2, 3, 4, 5, 6, 7, 8, 9, 10, 11, 12)    # time in months
t= sin(x*pi/6)
X=matrix(c(u,x,t), nrow=12, ncol=3)
Y=c(386.92, 387.41, 388.77, 389.46, 390.18, 389.43, 387.74,
   385.91, 384.77, 384.38, 385.99, 387.27)    # CO2 measurements
beta=solve(t(X)%*%X, t(X)%*%Y)
beta                                            # partial slopes
```

```
##            [,1]
## [1,] 386.9502
## [2,]   0.0619
## [3,]   2.1282
```

The model is: $y = 386.95016774 + 0.06189727x + 2.12824495\sin(\pi x/6)$

2. Plotting the points and the curve (see Figure 2.6).

```
plot(x,Y,  col = "blue", pch = 15, lwd = 2,
        main="Sinusoidal Model", col.main="darkgreen")
curve(386.95016774 + 0.06189727*x + 2.12824495*sin(pi*x/6),
            0, 13, add=TRUE, col = "red", lwd = 2)
legend('topright',inset=0.05,c("y = 386.9502+0.0619*x
    + 2.1282*sin(pi*x/6)"), lty=1,col=c("red"),title="best curve fit")
```

3. Test of the Validity of the Sinusoidal Model. We would like to check whether the sinusoidal model is a realistic description of the data. We test the hypothesis:

$$H_0 : \beta_2 = 0 \qquad \text{versus} \qquad H_a : \beta_2 \neq 0$$

Using matrix notations, one can write:

$$a = \begin{bmatrix} 0 \\ 0 \\ 1 \end{bmatrix} \qquad \beta = \begin{bmatrix} \beta_0 \\ \beta_1 \\ \beta_2 \end{bmatrix} \qquad \beta_2 = {}^t a.\beta$$

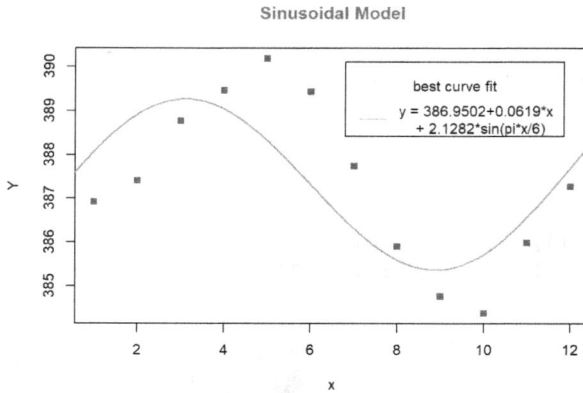

FIGURE 2.6 Sinusoidal Model

```
a=c(0,0,1)
beta2=t(a)%*%beta
beta2
```

```
##          [,1]
## [1,] 2.128
```

Test statistic $$T = \frac{{}^t a \widehat{\beta} - {}^t a \widehat{\beta_0}}{S \sqrt{{}^t a ({}^t X X)^{-1} a}} \quad \rightsquigarrow \quad \mathcal{T}_{n-(k+1)}$$

has a Student t-distribution in repeated sampling, with $df = n - (k+1)$ where

$$ {}^t a ({}^t X.X)^{-1} a = {}^t a \begin{bmatrix} c_{00} & c_{01} & c_{02} \\ c_{10} & c_{11} & c_{12} \\ c_{20} & c_{21} & c_{22} \end{bmatrix} a = c_{22} $$

$$ V({}^t a \widehat{\beta}) = {}^t a ({}^t X.X)^{-1} a \sigma^2 = c_{22} \sigma^2 $$

Calculating c_{22}

```
A=t(X)%*%X
I=matrix(c(1,0,0,0,1,0,0,0,1), nrow=3, ncol=3)
```

```
C=solve(A,I)
C                           # C is the inverse of A
```

```
##            [,1]      [,2]     [,3]
## [1,]   0.7942 -0.10937 -0.4082
## [2,]  -0.1094  0.01683  0.0628
## [3,]  -0.4082  0.06280  0.4010
```

```
c22=C[3,3]
c22
```

```
## [1] 0.401
```

An estimator of σ^2 is

$$S^2 = \frac{SSE}{n-(k+1)} \qquad \text{with} \qquad SSE = \sum (y_i - \widehat{y}_i)^2 = {}^t YY - {}^t \widehat{\beta}({}^t XY)$$

Here, we have

```
SSE = t(Y)%*%Y - (t(beta))%*%(t(X)%*%Y)
k=2
n=12
S2 = SSE/(n-(k+1))
S = sqrt(S2)
cat("The value of S is ", S)
```

```
## The value of S is  1.363
```

The observed statistic is given by $\qquad t = \dfrac{\widehat{\beta}_2 - \beta_2}{s\sqrt{c_{22}}}$

```
t=(beta2-0)/(S*sqrt(c22))
t
```

```
##          [,1]
## [1,] 2.465
```

► **The Rejection Region Approach.** With the level of significance $\alpha = 0.05$, the t-value:

$$t_{\alpha/2;df=n-(k+1)} = t_{0.025;12-3=9} = t_0$$

```
n=12
k=2
t0=qt(p=0.05/2, df=n-(k+1),lower.tail = FALSE, log.p = FALSE)
t0
```

```
## [1] 2.262
```

and the rejection region is characterized by $|t| > t_{\alpha/2} = 2.262157$.

The observed value t satisfies $|t| = 2.465217 > 2.262157 = t_{\alpha/2}$; that is, t falls in the rejection region. We can reject H_0. The data presents sufficient evidence to indicate that the model is sinusoidal.

► **The p-value Approach.** Because the test is two tailed, we have

$$p\text{-value} = 2P(|T| > t)$$

```
pv=2*pt(q=abs(t), df=n-(k+1), lower.tail = FALSE, log.p = FALSE)
pv
```

```
##            [,1]
## [1,] 0.03585
```

The p-value is less than the level of significance α. H_0 is rejected.

► **The Confidence Interval approach.**

A $100(1 - \alpha)\%$ confidence interval for ${}^t a\beta$ is

$${}^t a\widehat{\beta} \pm t_{\alpha/2} S \sqrt{{}^t a ({}^t X.X)^{-1} a}$$

```
LCB= beta2 - t0*S*sqrt(c22)
UCB= beta2 + t0*S*sqrt(c22)
cat("LCB =", LCB, "      ;  ", "UCB=", UCB)
```

```
## LCB = 0.1753        ;      UCB= 4.081
```

The calculated interval is $(0.1753034, 4.081186)$ *and it doesn't include the value 0; the value hypothesized in* H_0.

This indicates that it is quite possible that $\beta_2 \neq 0$, *and therefore agrees with the test conclusions with the Critical and p-value approaches: Reject* H_0: $\beta_2 = 0$.

We notice the agreement between the conclusions reached by the three approaches.

2.4 Predicting a Particular Value of Y in a Multi-linear Regression

Assume that we have fitted a multiple linear-regression model

$$Y = \beta_0 + \beta_1 x_1 + \beta_2 x_2 + \ldots + \beta_k x_k + \varepsilon$$

and that we are interested in predicting the value of Y^* when $x_1 = x_1^*$, $x_2 = x_2^*$, ..., $x_k = x_k^*$. We predict the value Y^* with

$$\widehat{Y^*} = \widehat{\beta_0} + \widehat{\beta_1} x_1^* + \widehat{\beta_2} x_2^* + \ldots + \widehat{\beta_k} x_k^* = {}^t a \widehat{\beta} \qquad \text{where} \qquad a = \begin{bmatrix} 1 \\ x_1^* \\ \vdots \\ x_k^* \end{bmatrix}.$$

We focus on the difference between the variable Y^* and the predicted value $\widehat{Y^*}$:

$$error = Y^* - \widehat{Y^*}.$$

Because both Y^* and $\widehat{Y^*}$ are normally distributed, the error is normally distributed, and we have

▶ **Properties**

$$E(error) \; = \; 0 \qquad \text{and} \qquad V(error) = [1 + {}^t a({}^t X.X)^{-1} a]\, \sigma^2.$$

$$Z = \frac{Y^* - \widehat{Y^*}}{\sigma \sqrt{1 + {}^t a({}^t X X)^{-1} a}} \qquad \rightsquigarrow \qquad \mathcal{N}(0,1)$$

has a standard normal distribution.

– If we substitute S to σ, then

$$T = \frac{Y^* - \widehat{Y^*}}{S\sqrt{1 + {}^t a ({}^t XX)^{-1} a}} \quad \rightsquigarrow \quad \mathcal{T}_{n-(k+1)}$$

has a Student's t-distribution.

▶ **Confidence Interval.** The random C.I. is characterized by

$$P\left(\left| \frac{Y^* - \widehat{Y^*}}{S\sqrt{1 + {}^t a ({}^t XX)^{-1} a}} \right| < t_{\alpha/2, n-(k+1)} \right) = 1 - \alpha$$

The $(1-\alpha)100\%$ prediction interval for Y when $x_1 = x_1^*$, $x_2 = x_2^*$, ..., $x_k = x_k^*$ is

$${}^t a \widehat{\beta} - t_{\alpha/2, n-(k+1)} \, S \, \sqrt{1 + {}^t a ({}^t XX)^{-1} a} \; < \; Y$$

$$< \; {}^t a \widehat{\beta} + t_{\alpha/2, n-(k+1)} \, S \, \sqrt{1 + {}^t a ({}^t XX)^{-1} a}$$

where ${}^t a = [1, x_1^*, x_2^*, \ldots, x_k^*]$.

Solved Problems

2.4.1   Parabolic Model

1. Fit a parabola to the data:

    (x, y): $(-2, 0)$, $(-1, 0)$, $(0, 1)$, $(1, 1)$, $(2, 3)$

2. Find a 90% confidence interval for y when $x = 1$.
3. Predict the particular value of y with $1 - \alpha = 0.90$ when $x = 2$.
4. Do the data present sufficient evidence to indicate that x^2 contributes information for the prediction of Y? (Test the Hypothesis $H_0 : \beta_2 = 0$, using $\alpha = 0.05$.)

Solution. 1. Finding the Parabolic Model.

```
X=matrix(c(1, 1, 1, 1, 1, -2, -1, 0, 1, 2, 4, 1, 0, 1, 4),
                nrow=5, ncol=3)
Y=c(0, 0, 1, 1, 3)                          # y-coordinates
beta=solve(t(X)%*%X, t(X)%*%Y)              # partial slopes
beta
```

```
##           [,1]
## [1,] 0.5714
## [2,] 0.7000
## [3,] 0.2143
```

The model is described by the equation:

$$y = 0.5714286 + 0.7x + 0.2142857x^2.$$

2. Finding a 90% confidence interval for y when $x = 1$.

$$E(Y) = \beta_0 + \beta_1 x + \beta_2 x^2 =^t a\beta \qquad \beta = \begin{bmatrix} \beta_0 \\ \beta_1 \\ \beta_2 \end{bmatrix} \qquad a = \begin{bmatrix} 1 \\ x \\ x^2 \end{bmatrix}_{|x=1} = \begin{bmatrix} 1 \\ 1 \\ 1 \end{bmatrix}$$

The confidence interval is given by

$$^t a\widehat{\beta} \pm t_{\alpha/2} S \sqrt{1 +^t a(^t X.X)^{-1} a}.$$

An estimator of σ^2 is

$$S^2 = \frac{SSE}{n - (k+1)} \qquad \text{with} \qquad SSE = \sum (y_i - \widehat{y}_i)^2 =^t YY -^t \widehat{\beta}(^t XY)$$

```
SSE=t(Y)%*%Y - (t(beta))%*%(t(X)%*%Y )

k=2
n=5
S2= SSE/(n-(k+1))
S= sqrt(S2)
cat("an estimate of the standard deviation =", S)

## an estimate of the standard deviation = 0.4781
```

The matrix $C = (^t X.X)^{-1}$. Using matrix notation, we can write:

$$^t a(^t X.X)^{-1} a =^t a \begin{bmatrix} c_{00} & c_0 & c_{02} \\ c_{10} & c_{11} & c_{12} \\ c_{20} & c_{21} & c_{22} \end{bmatrix} a =^t aCa$$

```
A=t(X)%*%X
I=matrix(c(1, 0, 0, 0, 1, 0, 0, 0, 1), nrow=3, ncol=3)
C=solve(A,I)
C

##            [,1] [,2]     [,3]
## [1,]    0.4857  0.0 -0.14286
## [2,]    0.0000  0.1  0.00000
## [3,]   -0.1429  0.0  0.07143
```

The Confidence Interval for y when $x = 1$

```
a=c(1,1,1)
t1=qt(p=0.1/2, df=n-(k+1),lower.tail = FALSE, log.p = FALSE)

c0=t(a)%*%C%*%a

LCB = t(a)%*%beta - t1*S*sqrt(1+c0)
UCB = t(a)%*%beta + t1*S*sqrt(1+c0)
cat("LCB =", LCB, "        ;    ", "UCB=", UCB)


## LCB = -0.1491       ;     UCB= 3.121
```

The 90% *C.I* for the mean of y is given by: $-0.1491371 < y < 3.120566$.

3. Predicting the particular value of y **with** $1 - \alpha = 0.90$ **when** $x = 2$.

The confidence interval is given by

$$^t a \hat{\beta} \pm t_{\alpha/2} S \sqrt{1 +^t a (^t X.X)^{-1} a} \qquad\qquad a = \begin{bmatrix} 1 \\ x \\ x^2 \end{bmatrix}_{|x=2} = \begin{bmatrix} 1 \\ 2 \\ 4 \end{bmatrix}$$

```
x=2
a=c(1,x,x^2)
n=5
k=2
t1=qt(p=0.1/2, df=n-(k+1),lower.tail = FALSE, log.p = FALSE)
c0=t(a)%*%C%*%a

LCB= t(a)%*%beta - t1*S*sqrt(1+c0)
UCB= t(a)%*%beta + t1*S*sqrt(1+c0)
cat("LCB =", LCB, "        ;    ", "UCB=", UCB)


## LCB = 0.9115       ;     UCB= 4.746
```

The 90% *C.I* for the value of y is given by: $0.9115383 < y < 4.745605$.

■ Other R codes for reading a Prediction Interval

We apply the "lm" function to y as a function of x and x^2, and save the linear regression model in a new variable "fit".

```
x=c(-2,-1, 0, 1,2)
xx=x*x
y=c(0,0,1,1,3)
data1=data.frame(x,xx)
fit = lm(y ~ x + xx, data=data1)
```

For a given value of the independent variable x, the interval estimate of the dependent variable y is called the prediction interval. We can read the predicted value y and this interval by using the function "predict":

```
newdata=data.frame(x=2, xx=4)                    # wrap the parameter
predict(fit,newdata,level = 0.90,interval="predict")
```

```
##      fit    lwr    upr
## 1 2.829 0.9115 4.746
```

4. To explore whether x^2 contributes information for the prediction of Y, we test the Hypothesis $H_0 : \beta_2 = 0$ using $\alpha = 0.05$.

With the function "summary( )", one can access the significance of the coefficients:

```
summary(fit)
```

```
##
## Call:
## lm(formula = y ~ x + xx, data = data1)
##
## Residuals:
##        1       2       3       4       5
## -0.0286 -0.0857  0.4286 -0.4857  0.1714
##
## Coefficients:
##             Estimate Std. Error t value Pr(>|t|)
## (Intercept)    0.571      0.333    1.71    0.228
## x              0.700      0.151    4.63    0.044 *
## xx             0.214      0.128    1.68    0.236
## ---
```

```
## Signif. codes:
## 0 '***' 0.001 '**' 0.01 '*' 0.05 '.' 0.1 ' ' 1
##
## Residual standard error: 0.478 on 2 degrees of freedom
## Multiple R-squared:  0.924,  Adjusted R-squared:  0.848
## F-statistic: 12.1 on 2 and 2 DF,  p-value: 0.0762
```

The p-value of the test $H_0 : \beta_2 = 0$ is equal to 0.2355. It is greater than the level of significance α. H_0 is not rejected.

Unless we are willing to work with a relatively large value of α (at least 0.2355), we cannot reject H_0. This means that the variable x^2 is not contributing important information to the model.

2.4.2   Three Variables in a Multi-linear Model

A response Y is a function of three independent variables x_1, x_2, and x_3 that are related as follows:

$$Y = \beta_0 + \beta_1 x_1 + \beta_2 x_2 + \beta_3 x_3 + \varepsilon.$$

a. Fit this model to the following $n = 7$ data points:

$$(y, x_1, x_2, x_3) = (1, -3, 5, -1), \quad (0, -2, 0, 1), \quad (0, -1, -3, 1),$$
$$(1, 0, -4, 0), \quad (2, 1, -3, -1), \quad (3, 2, 0, -1), \quad (3, 3, 5, 1).$$

b. Predict Y when $x_1 = 1$, $x_2 = -3$, and $x_3 = -1$. Compare with the observed response in the original data. Why are these two not equal?

c. Do the data present sufficient evidence to indicate that x_3 contributes information for the prediction of Y? (Test the Hypothesis $H_0 : \beta_3 = 0$, using $\alpha = 0.05$.)

d. Find a 95% confidence interval for the expected value of Y, given $x_1 = 1$, $x_2 = -3$, and $x_3 = -1$.

e. Find a 95% prediction interval for Y, given $x_1 = 1$, $x_2 = -3$, and $x_3 = -1$.

(This problem is formulated from D.D. Wackerly (2008) Exercise 11.97.)

Solution.

a. Fitting the multi-linear model to the $n = 7$ data points.

▶ We apply the "lm" function to y as a function of x_1, x_2, and x_3, and save the linear regression model in a new variable "fit". Then we read the parameters of the estimated regression equation by calling "fit".

```
x1=c(-3, -2, -1, 0, 1, 2, 3)
x2=c(5, 0, -3, -4, -3, 0, 5)
x3=c(-1, 1, 1, 0, -1, -1, 1)
y =c(1, 0, 0, 1, 2, 3, 3)
```

```
DF=data.frame(x1,x2,x3)
model= y ~ x1 + x2 + x3          # Store the model to estimate
fit = lm(model, data=DF)
fit                              # Estimate the model with the data frame DF
```

```
##
## Call:
## lm(formula = model, data = DF)
##
## Coefficients:
## (Intercept)           x1            x2            x3
##       1.429        0.500         0.119        -0.500
```

The multi-linear model is

$$y = 1.429 + 0.500x_1 + 0.119x_2 - 0.500x_3.$$

▶ With the function "summary( )", one can access more details such as the significance of the coefficients, the degree of freedom, and the residuals.

```
summary(fit)
```

```
##
## Call:
## lm(formula = model, data = DF)
##
## Residuals:
##        1        2        3        4        5        6
## -0.0238   0.0714  -0.0714   0.0476  -0.0714   0.0714
##        7
## -0.0238
##
## Coefficients:
##                Estimate Std. Error t value Pr(>|t|)
## (Intercept)    1.42857    0.03367     42.4  2.9e-05 ***
## x1             0.50000    0.01684     29.7  8.4e-05 ***
## x2             0.11905    0.00972     12.2  0.00117 **
## x3            -0.50000    0.03637    -13.8  0.00083 ***
## ---
## Signif. codes:
## 0 '***' 0.001 '**' 0.01 '*' 0.05 '.' 0.1 ' ' 1
##
```

```
## Residual standard error: 0.0891 on 3 degrees of freedom
## Multiple R-squared:  0.998,  Adjusted R-squared:  0.995
## F-statistic:  407 on 3 and 3 DF,  p-value: 0.000206
```

From the value of the adjusted R-squared, the model explains 99.51% of the variance in Y.

b. Predicting Y when $x_1 = 1, x_2 = -3$, and $x_3 = -1$.

```
newdata=data.frame(x1=1, x2=-3, x3=-1)    #wrap the parameter
predict(fit,newdata)
```

```
##     1
## 2.071
```

The predicted value 2.071429 is different from the observed value 2. The predicted value is obtained using the approximation model obtained by the Least Squares Method. The reliability of this approximation can be estimated by finding a confidence interval.

c. To explore whether x_3 contributes information for the prediction of Y, we test the Hypothesis $H_0 : \beta_3 = 0$, using $\alpha = 0.05$.

From the "summary( )", the p-value $= 0.000833$ in the test $\beta_3 = 0$ is less than $\alpha = 0.05$. Therefore, the variable x_3 is statistically significant in the multiple linear regression model. It is not possible to have $\beta_3 = 0$. One can read from the summary similar conclusions regarding the contributions of x_1 and x_2 in the model.

▶ Inference from the "summary(fit)" table output:

– The "summary(fit)" table proves that there is a strong negative relationship between x_3 and y and a positive relationship with x_1 and x_2.

   – All three variables x_1, x_2, x_3 have a statistical impact on y. Indeed, each of the tests

$$H_0 : \beta_i = 0 \quad \text{(No statistical impact)} \qquad\qquad \text{versus}$$

$$H_a : \beta_i \neq 0 \quad \text{(The predictor has a meaningful impact on y)} \quad i = 1, 2, 3$$

shows a p-value < 0.001. This indicates that the variable x_i is statistically significant.

▶ We run the ANOVA F-test that shows a p-value $= 0.0002058$ less than 0.001. We conclude that the regression is highly significant.

```
Model0 = lm(y ~ 1, DF)
anova(Model0, fit)
```

```
## Analysis of Variance Table
##
## Model 1: y ~ 1
## Model 2: y ~ x1 + x2 + x3
##   Res.Df  RSS Df Sum of Sq    F Pr(>F)
## 1      6 9.71
## 2      3 0.02  3      9.69  407 0.00021 ***
## ---
## Signif. codes:
## 0 '***' 0.001 '**' 0.01 '*' 0.05 '.' 0.1 ' ' 1
```

d. Finding a 95% confidence interval for the expected value of $E(Y)$, given $x_1 = 1$, $x_2 = -3$, and $x_3 = -1$.

We have

$$E(Y) = \beta_0 + \beta_1 x_1 + \beta_2 x_2 + \beta_3 x_3 = {}^t a\beta \qquad\qquad a = \begin{bmatrix} 1 \\ x_1 \\ x_2 \\ x_3 \end{bmatrix} = \begin{bmatrix} 1 \\ 1 \\ -3 \\ -1 \end{bmatrix}$$

The confidence interval is given by

$$ {}^t a\widehat{\beta} \pm t_{\alpha/2} S \sqrt{{}^t a ({}^t X.X)^{-1} a}.$$

An estimator of σ^2 is

$$S^2 = \frac{SSE}{n - (k+1)} \qquad \text{with} \qquad SSE = \sum (y_i - \widehat{y_i})^2 = {}^t YY - {}^t \widehat{\beta}({}^t XY)$$

```
x0=c(1,1,1,1,1,1,1)
X= matrix(c(x0,x1,x2,x3),nrow=7, ncol=4)
Y= y
beta=solve(t(X)%*%X, t(X)%*%Y)          # partial slopes
beta
```

```
##          [,1]
## [1,]   1.429
## [2,]   0.500
## [3,]   0.119
## [4,]  -0.500
```

```
SSE=t(Y)%*%Y - (t(beta))%*%(t(X)%*%Y )
```

```
k=3
n=7
S2= SSE/(n-(k+1))
S= sqrt(S2)
cat("The sample standard deviation = ", S)
```

```
## The sample standard deviation =  0.08909
```

The matrix $C = ({}^t X.X)^{-1}$. Using matrix notation, we have:

$$ {}^t a ({}^t X.X)^{-1} a = {}^t a \begin{bmatrix} c_{00} & c_{01} & c_{02} & c_{03} \\ c_{10} & c_{11} & c_{12} & c_{13} \\ c_{20} & c_{21} & c_{22} & c_{23} \\ c_{30} & c_{31} & c_{32} & c_{33} \end{bmatrix} a = {}^t a C a $$

```
A=t(X)%*%X
A
```

```
##      [,1] [,2] [,3] [,4]
## [1,]    7    0    0    0
## [2,]    0   28    0    0
## [3,]    0    0   84    0
## [4,]    0    0    0    6
```

```
I=matrix(c(1,0,0,0,0,1,0,0,0,0,1,0,0,0,0,1), nrow=4, ncol=4)
C=solve(A,I)
C
```

```
##          [,1]     [,2]    [,3]    [,4]
## [1,]  0.1429 0.00000 0.0000 0.0000
## [2,]  0.0000 0.03571 0.0000 0.0000
## [3,]  0.0000 0.00000 0.0119 0.0000
## [4,]  0.0000 0.00000 0.0000 0.1667
```

The Confidence Interval for $E(Y)$ when $(x_1, x_2, x_3) = (1, -3, -1)$

```
k=3
n=7
a=c(1,1,-3,-1)
c0=t(a)%*%C%*%a
```

```
t1=qt(p=0.05/2, df=n-(k+1),lower.tail = FALSE, log.p = FALSE)
```

```
LCB = t(a)%*%beta - t1*S*sqrt(c0)
UCB = t(a)%*%beta + t1*S*sqrt(c0)
cat("LCB =", LCB, "      ;   ", "UCB=", UCB)
```

```
## LCB = 1.881      ;     UCB= 2.262
```

The 95% confidence interval for $E(Y)$ is: $1.880739 < E(Y) < 2.262119$.

e. Find a 95% prediction interval for Y, given $x_1 = 1$, $x_2 = -3$, and $x_3 = -1$.

For a given x, the interval estimate of the dependent variable y is called the prediction interval. We can read the predicted value y and this interval by using the function "predict":

```
newdata=data.frame(x1=1, x2=-3, x3=-1)       # wrap the parameter
predict(fit,newdata,level = 0.95,interval="predict")
```

```
##      fit  lwr   upr
## 1 2.071 1.73 2.413
```

The C.I. is: $1.729751 < y < 2.413106$. It can be obtained also as follows:

```
LCB1 = t(a)%*%beta - t1*S*sqrt(1+c0)
UCB1 = t(a)%*%beta + t1*S*sqrt(1+c0)
cat("Lower bound =", LCB1, "    ;    ", "Upper bound =", UCB1)
```

```
## Lower bound = 1.73       ;      Upper bound = 2.413
```

2.5   A Test for H_0: $\beta_{g+1} = \beta_{g+2} = \ldots = \beta_k = 0$

We want to fit a model involving only a subset of the independent variables $X_1, X_2, \ldots, X_r$; that is, fit a **reduced** model R to the data:

Model R.

$$y = \beta_0 + \beta_1 x_1 + \beta_2 x_2 + \ldots + \beta_r x_r$$

and find a relation between model R and model C, the **complete** model:

Model C.

$$y = \beta_0 + \beta_1 x_1 + \beta_2 x_2 + \ldots + \beta_r x_r + \beta_{r+1} x_{r+1} + \ldots + \beta_k x_k$$

▶ **Test.** In order to compare the two models, we test the following null Hypothesis H_0 versus the alternative hypothesis H_a:

$$H_0: \quad \beta_{r+1} = \beta_{r+2} = \ldots = \beta_k = 0 \qquad \text{versus}$$

$$H_a: \quad \text{at least one of the parameters } \beta_{r+1}, \ldots, \beta_k \text{ differs from 0.}$$

▶ **Statistic.** Assume H_0 is true. Then, if SSE_R and SSE_C denote the sum of squares of errors for the reduced model and the complete model, respectively, we have

$$\frac{SSE_R}{\sigma^2} = \frac{SSE_C}{\sigma^2} + \frac{(SSE_R - SSE_C)}{\sigma^2}.$$

Each of these random variables has a χ^2 distribution;

$$\frac{SSE_R}{\sigma^2} \rightsquigarrow \chi^2_{n-(r+1)}, \qquad \frac{SSE_C}{\sigma^2} \rightsquigarrow \chi^2_{n-(k+1)},$$

$$\frac{(SSE_R - SSE_C)}{\sigma^2} \rightsquigarrow \chi^2_{k-r}.$$

Consequently, the statistic for this test is the random variable F defined below with an $\mathcal{F}$ distribution;

$$F = \frac{\frac{(SSE_R - SSE_C)/\sigma^2}{k - r}}{\frac{SSE_C/\sigma^2}{n - (k+1)}} = \frac{(SSE_R - SSE_C)/(k - r)}{SSE_C/[n - (k+1)]} \rightsquigarrow \mathcal{F}_{df_1 = k - r \ ; \ df_2 = n - (k+1)}$$

▶ **Rejection Region**

With a level of significance α, the null Hypothesis H_0 is rejected if $F > F_\alpha$.

$$\alpha = P\Big(H_0 \text{ is rejected } \Big| H_0 \text{ is true }\Big) = P\Big(F > F_\alpha\Big)$$

This will occur when $(SSE_R - SSE_C)$ is large, leading to F to be large too. In particular when SSE_C is small; that is, the variables $x_{r+1}, \ldots, x_k$, in model C, contribute more for the prediction of y than model R.

In fact, the greater the difference between SSE_R and SSE_C, the greater is the evidence to indicate that model C contributes more information for the prediction of y than model R. (William Mendenhall, 2018)

▶ **Case** $r = 0$: In this case, **the reduced model** is $y = \beta_0 = \bar{y}$ and we test the null Hypothesis H_0 versus the alternative Hypothesis H_a:

$$H_0: \quad \beta_1 = \beta_2 = \ldots = \beta_k = 0 \qquad \text{versus}$$

$$H_a: \quad \text{at least one of the parameters } \beta_1, \beta_2, \ldots, \beta_k \text{ differs from 0.}$$

Then

$$F = \frac{(SSE_R - SSE_C)/k}{(SSE_C)/(n - (k+1))} = \frac{\text{mean square (model)}}{\text{mean square(error)}} \rightsquigarrow \mathcal{F}_{df_1 = k \ ; \ df_2 = n - (k+1)}$$

which can be written as

$$F = \frac{n - (k+1)}{k}\left(\frac{R^2}{1 - R^2}\right)$$

where R^2 is the multiple coefficient of determination defined by:

$$R^2 = \frac{S_{yy} - SSE_C}{S_{yy}} \qquad \text{with} \qquad SSE_R = S_{yy}.$$

In the denominator of R^2, S_{yy} quantifies the variation in the y-values.

In the numerator of R^2, $S_{yy} - SSE_C$ quantifies the amount of variation in the y's that is explained by the complete set of independent variables $x_1, x_2, \ldots, x_k$.

Therefore, R^2 represents the proportion of the total sample variation in y that can be explained by the multiple-regression model.

The value of R^2 can never decrease with the addition of more variables into the regression model. Hence, R^2 can be inflated by the inclusion of more and more predictor variables. An alternative measure of the strength of the regression model is adjusted for degrees of freedom by using mean squares rather than sums of squares.

As an alternative to R^2 as a measure of adequacy, the adjusted multiple coefficient of determination, R^2_{adj} is reported. It is defined by

$$R^2_{adj} = \frac{S_{yy}/(n-1) - MSE}{S_{yy}/(n-1)} \qquad \text{where} \qquad MSE = \frac{SSE_C}{n - (k+1)}$$

which we can write

$$R^2_{adj} = 1 - \left[\frac{n-1}{n-(k+1)}\right]\left(\frac{SSE_C}{S_{yy}}\right) = 1 - \left[\frac{n-1}{n-(k+1)}\right](1 - R^2)$$

Note that $R^2_{adj} \leqslant R^2$.

▶ **Remark.**

We use the analysis of variance F test and R^2 to determine how well the model fits the data.

Solved Problems

2.5.1   Parabolic Model

Consider the data:

$$(x, y): \quad (-2, 0), \quad (-1, 0), \quad (0, 1), \quad (1, 1), \quad (2, 3)$$

Do the data provide sufficient evidence to indicate that the second order model contributes information for the prediction of Y? Use $\alpha = 0.05$.

Solution. We test the hypotheses

$$H_0 : \beta_1 = \beta_2 = 0 \qquad \text{versus}$$

$$H_a : \text{ at least one of the parameters } \beta_1, \beta_2, \text{ differs from } 0.$$

▶ **The complete model** is described by the equation

$$Y = \beta_0 + \beta_1 x + \beta_2 x^2 + \varepsilon$$

```
X=matrix(c(1,1,1,1,1, -2,-1,0,1,2, 4,1,0,1,4), nrow=5, ncol=3)
Y=c(0, 0, 1, 1, 3)                    # y-coordinates
beta=solve(t(X)%*%X, t(X)%*%Y)
beta                                  # partial slopes
```

```
##          [,1]
## [1,]  0.5714
## [2,]  0.7000
## [3,]  0.2143
```

The model is illustrated by the parabola: $y = 0.5714286 + 0.7x + 0.2142857x^2$ in Figure 2.7.

▶ **The reduced model** is described by the equation:

$$Y = \beta_0 + \varepsilon$$

where $\beta_0 = 1$. Indeed, we have

```
RX=matrix(c(1,1,1,1,1), nrow=5, ncol=1)
Y=c(0, 0, 1, 1, 3)
Rbeta=solve(t(RX)%*%RX, t(RX)%*%Y)
Rbeta                    # partial slopes for the reduced model
```

```
##        [,1]
## [1,]     1
```

The horizontal line in Figure 2.7 illustrates this approximation.

```
Xp=c(-2, -1, 0, 1, 2)
Y=c(0, 0, 1, 1, 3)
plot(Xp,Y,  col = "blue", pch = 15)
curve(0.5714286 + 0.7*x + 0.2142857*x^2, -2, 2, add=TRUE,
      col = "red", lwd = 2)
curve(1+x-x, -2, 2, add=TRUE, col = "darkgreen", lwd = 2)
legend('topleft',inset=0.08, c("Complete Model","Reduced Model"),
      lty=1,col=c("red", "darkgreen") )
```

▶ **Comparing the two models: The rejection region approach**

The statistic is:

$$F = \frac{(SSE_R - SSE_C)/(k-g)}{(SSE_C)/(n-(k+1))} \quad \rightsquigarrow \quad \mathcal{F}_{df_1=k-g;df_2=n-(k+1)}$$

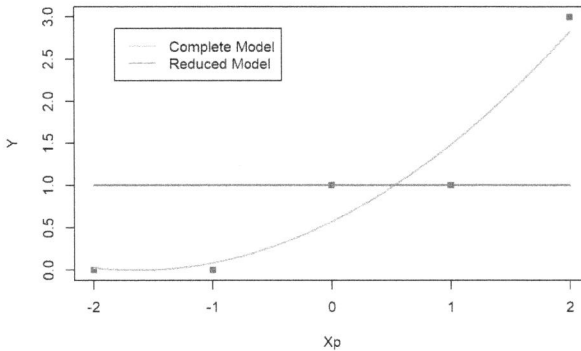

FIGURE 2.7 Reduced and complete models

```
g=0        # number of variables in the Reduced Model
k=2        # number of variables in the Complete Model
n=5        # sample size

SSE_R = t(Y)%*%Y - (t(Rbeta))%*%(t(RX)%*%Y)
SSE_C = t(Y)%*%Y - (t(beta))%*%(t(X)%*%Y)

F = ((SSE_R - SSE_C)/(k-g))/((SSE_C)/(n-(k+1)))
F


##         [,1]
## [1,] 12.12

alpha=0.05                              # Level of significance
a= qf(alpha, df1=k-g, df2=n-(k+1), lower.tail = FALSE)
a                                       # a=F_{alpha; df1; df2}


## [1] 19
```

The observed value $F = 12.125$ is less than $F_\alpha = 19$. Hence, it does not fall in the rejection region $[F > F_\alpha]$. We conclude that, at the level of significance $\alpha = 0.05$, there is not enough evidence to support a claim that either β_1 or β_2 differs from 0.

▶ **Comparing the two models: The p-value approach**

The p-value of the F test is calculated by:

$$p\text{-value} = P(F > a)$$

```
pv=pf(q=F,df1=k-g, df2=n-(k+1), lower.tail = FALSE, log.p = FALSE)
pv
```

```
##             [,1]
## [1,] 0.07619
```

The p-value is greater than the level of significance α. We cannot reject H_0. This agrees with the critical region approach.

2.5.2   Reliability of a Conic Surface Model

It is desired to relate abrasion resistance of rubber (Y) to the amount of silica filler x_1 and the amount of coupling agent x_2.

Fine particle silica fibers are added to rubber to increase strength and resistance to abrasion. The coupling agent chemically bonds the filler to the rubber polymer chains and thus increases the efficiency of the filler (see Table 2.2).

    a. Fit the second order model

$$y = \beta_0 + \beta_1 x_1 + \beta_2 x_2 + \beta_3 x_1^2 + \beta_4 x_2^2 + \beta_5 x_1 x_2 + \varepsilon$$

to the data.

    b. Test $H_0 : \beta_3 = \beta_4 = \beta_5 = 0$. Indicate the proper conclusion if we choose $\alpha = 0.05$.

We are testing that the surface is actually a plane versus the alternative that it is a conic surface.

(This problem is formulated from D.D. Wackerly (2008) Exercise 11.19)

Solution. a. Fitting the second order model to the data.

▶ We apply the "lm" function to y as a function of x_1, x_2, x_1^2, x_2^2 and $x_1 x_2$, and save the linear regression model in a new variable "Model_C". Then we read the parameters of the estimated regression equation by calling "Model_C".

```
x1=c(1,1,-1,-1,0,0,0,0,0,1.5,-1.5)     # silica filler amount
x2=c(-1,1,1,-1,0,0,0,1.5,-1.5,0,0)     # coupling agent amount
x1x1=x1*x1
x2x2=x2*x2
```

```
x1x2=x1*x2

# abrasion resistance of rubber
y=c(83, 113, 92, 82, 100, 96, 98, 95, 80, 100, 92)
DF=data.frame(y,x1,x2, x1x1, x2x2, x1x2)

# Estimate the complete model
Model_C = lm(y ~ x1 + x2 +x1x1 + x2x2 + x1x2, data=DF)
```

▶ With the function "summary()", we can access more details such as the significance of the coefficients, the values of R^2 and F, and the shape of the residuals.

```
summary(Model_C)
```

```
##
## Call:
## lm(formula = y ~ x1 + x2 + x1x1 + x2x2 + x1x2, data = DF)
##
## Residuals:
##         1        2        3        4        5        6
## -1.11631  4.17781  1.17781 -4.11631  1.99544 -2.00456
##         7        8        9       10       11
## -0.00456 -3.55675  3.50208 -2.02733  1.97267
##
## Coefficients:
##              Estimate Std. Error t value Pr(>|t|)
## (Intercept)   98.005      2.269   43.19  1.3e-07 ***
## x1             4.000      1.354    2.95   0.0318 *
## x2             7.353      1.354    5.43   0.0029 **
## x1x1          -0.879      1.527   -0.58   0.5899
## x2x2          -4.657      1.527   -3.05   0.0285 *
## x1x2           5.000      1.974    2.53   0.0524 .
## ---
## Signif. codes:
## 0 '***' 0.001 '**' 0.01 '*' 0.05 '.' 0.1 ' ' 1
##
## Residual standard error: 3.95 on 5 degrees of freedom
## Multiple R-squared:  0.915,  Adjusted R-squared:  0.831
## F-statistic: 10.8 on 5 and 5 DF,  p-value: 0.0103
```

► **Inference from the above outputs:**

    • The above table proves that there is a strong negative relationship between x_2^2 and y ($\beta_4 = -4.6565$) and a strong positive relationship with x_1 and x_2 ($\beta_5 = 5.0000$).

    • Only the three variables x_1, x_2, x_2^2 have a statistical impact on y. Indeed, each of the tests

$$H_0 : \beta_i = 0 \quad \text{(No statistical impact)} \quad\quad\quad \text{versus}$$

$$H_a : \beta_i \neq 0 \quad \text{(The predictor has a meaningful impact on y)} \quad i = 1, 2, 4$$

shows a p-value equal to $0.03175, 0.00287, 0.02846$ respectively, which is less than 0.05. This indicates that the variables x_1, x_2, and x_2^2 are statistically significant.

    • Adjusted R-squared; the variance explained by the model shows that the model explained 83.09% of the variance of y.

► **We run the ANOVA F-test.** The p-value $= 0.01028$ is less than 0.05. We conclude that the regression is highly significant. The comparison below using "anova(Model0, Model_C)" allows the reading of the sum of squares of the complete model. We can also read it using "anova(Model_C)".

```
Model0 = lm(y ~ 1, DF)
anova(Model0, Model_C)

## Analysis of Variance Table
##
## Model 1: y ~ 1
## Model 2: y ~ x1 + x2 + x1x1 + x2x2 + x1x2
##    Res.Df RSS Df Sum of Sq     F Pr(>F)
## 1      10 922
## 2       5  78  5       844  10.8   0.01 *
## ---
## Signif. codes:
## 0 '***' 0.001 '**' 0.01 '*' 0.05 '.' 0.1 ' ' 1
```

b. Testing $H_0 : \beta_3 = \beta_4 = \beta_5 = 0$ with $\alpha = 0.05$.

▶ **Find the reduced model**

```
Model_R = lm(y ~ x1 + x2, data=DF)
```

▶ **Significance of the coefficients.**

```
summary(Model_R)
```

```
##
## Call:
## lm(formula = y ~ x1 + x2, data = DF)
##
## Residuals:
##     Min     1Q Median     3Q    Max
## -9.757 -3.889  0.273  4.273  7.920
##
## Coefficients:
##             Estimate Std. Error t value Pr(>|t|)
## (Intercept)    93.73       1.93   48.65  3.5e-11 ***
## x1              4.00       2.19    1.83     0.11
## x2              7.35       2.19    3.35     0.01 *
## ---
## Signif. codes:
## 0 '***' 0.001 '**' 0.01 '*' 0.05 '.' 0.1 ' ' 1
##
## Residual standard error: 6.39 on 8 degrees of freedom
## Multiple R-squared:  0.646,  Adjusted R-squared:  0.557
## F-statistic: 7.29 on 2 and 8 DF,  p-value: 0.0157
```

The p-value $= 0.105$ shows that the contribution of the variable x_1 is not statistically significant.

▶ **ANOVA** F**-test for the Reduced Model**

```
Model0 = lm(y ~ 1, DF)
anova(Model0, Model_R)


## Analysis of Variance Table
##
## Model 1: y ~ 1
## Model 2: y ~ x1 + x2
##   Res.Df RSS Df Sum of Sq    F Pr(>F)
## 1     10 922
## 2      8 327  2       596 7.29  0.016 *
## ---
## Signif. codes:
## 0 '***' 0.001 '**' 0.01 '*' 0.05 '.' 0.1 ' ' 1
```

From the p-value $= 0.01574$ of the F-test, the reduced model fits the data well.

▶ Comparing the Reduced and the Complete Models

We perform an F-test for this purpose by computing the statistic F as follows:

```
n=11
k=5
r=2
SSE_R = 326.62                    # read from anova(Model0, Model_R)
SSE_C = 77.95                     # read from anova(Model0, Model_C)
F=((SSE_R - SSE_C)/(k-r))/(SSE_C/(n-(k+1)) )
F
```

```
## [1] 5.317
```

The critical value F_α is

```
alpha=0.05                        #a=F_{alpha; df1; df2}
a= qf(alpha, df1=k-r, df2=n-(k+1), lower.tail = FALSE)
cat("F- critical value =", a)
```

```
## F- critical value = 5.409
```

The observed value $F = 5.31687$ is less than $F_\alpha = 5.409451$. Hence, it does not fall in the rejection region $[F > F_\alpha]$. We cannot reject $H_0 : \beta_3 = \beta_4 = \beta_5 = 0$. We conclude that, at the level of significance $\alpha = 0.05$, there is not enough evidence to support a claim that the second order model fits the data significantly better than does the planar model.

The p-value of the test is given by

```
pf(F,df1=k-r, df2=n-(k+1), lower.tail = FALSE)
```

```
## [1] 0.0516
```

and is greater than the level of significance α. This agrees with the Rejection region conclusion.

Note that we have tested whether the group of variables x_1^2, x_2^2, and x_1x_2 contributed to a significantly better fit of the model to the data.

Hence, the planar model explains the abrasion resistance of rubber y more effectively than having to use measurements on the second order terms.

Using "Coefplot", confidence intervals for the coefficients β_i are plotted for the two models. The three confidence intervals for the reduced model do not contain the zero value (see Figure 2.8).

```
require(coefplot)
```

```
## Loading required package: coefplot
```

```
multiplot(Model_C, Model_R )
```

R codes for comparing the two models

```
anova(Model_R, Model_C)
```

```
## Analysis of Variance Table
##
## Model 1: y ~ x1 + x2
## Model 2: y ~ x1 + x2 + x1x1 + x2x2 + x1x2
##    Res.Df RSS Df Sum of Sq    F Pr(>F)
## 1       8 327
```

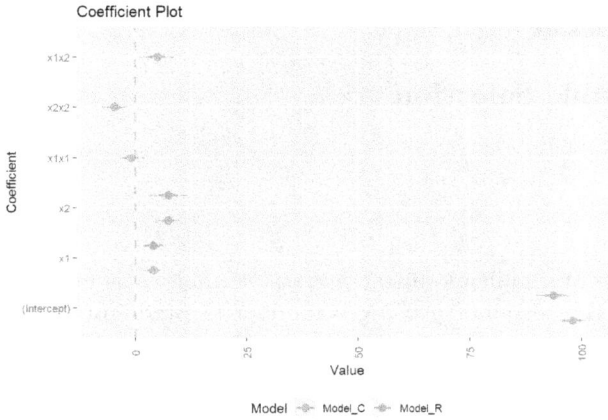

FIGURE 2.8 Confidence intervals: Reduced and Complete Models

```
## 2      5  78  3       249 5.32   0.052 .
## ---
## Signif. codes:
## 0 '***' 0.001 '**' 0.01 '*' 0.05 '.' 0.1 ' ' 1
```

From "the"anova(Model_R, Model_C)", we can read the value of the F statistic for the comparison of the two models, from which we can deduce the p-value $= 0.0516$. We can also read the sum of squares of errors for each model.

2.6   Variable Selection

One objective of a multiple linear regression analysis is to produce the best estimator of the response y using a number of predictor variables. With k variables, there are

$$C_k^1 + C_k^2 + \ldots + C_k^k = (1+1)^k - C_k^0 = 2^k - 1$$

possible models to be fitted. Thus, finding the best number k of variables for a model has several benefits in describing the model in a simpler way by avoiding unnecessary variables and minimizing the cost and time of extra measurements.

Stepwise procedures bring answers to this task. We illustrate the "Stepwise Procedure-Backward Elimination" in one problem by following the scheme:

1.   Start with the complete model, including all of the predictors.

2.   Test the usefulness of the model through the:

    − Analysis of Variance F-test

    − Significance of coefficients

    − R^2 and R_{adj}^2.

3.   Remove the predictor with highest p-value greater than a certain level of significance α.

4.   Refit the model with the remaining predictors.

5.   Compare the new model to the previous one using an F test and repeat steps $1, 2, 3$, and 4.

6.   Stop when all p-values are less than α.

Two other stepwise procedures are the "Forward Selection", where the backward method is reversed, and the "Stepwise Regression", where a combination of backward elimination and forward selection is adopted.

These methods are easily implemented in computers. However, they don't ensure the optimal choice of variables, as the criterion of p-values tends to reduce the number of variables that might be better to keep for predictive purposes (S. Dowdy, 2004).

Solved Problems - mtcars

Illustrate the backward method on the data set "mtcars" for the prediction of the variable "mpg" with respect to the five numerical variables "disp", "hp", "drat", "wt", "qsec". Use $\alpha = 0.05$. Which predictors should be included—can you tell from the p-values?

Solution. We illustrate the backward method. At each stage, we remove the predictor with the largest p-value over $\alpha = 0.05$ and make other statistical testings.

Step 1: The complete model = Model 1

Equation of Model 1

First, we read the data from R-Help. "mtcars" is data extracted from the 1974 Motor Trend Car Road Tests US magazine and comprises fuel consumption and 10 aspects of automobile design and performance for 32 automobiles (1973-74 models). The data contains 32 observations on the 11 variables:

mpg	Miles/(US) gallon
cyl	Number of cylinders
disp	Displacement (cu.in.)
hp	Gross horsepower
drat	Rear axle ratio
wt	Weight (1000 lbs)
qsec	1/4 mile time
vs	Engine (0 = V-shaped, 1 = straight)
am	Transmission (0 = automatic, 1 = manual)
gear	Number of forward gears
carb	Number of carburetors

```
library(ggplot2)
dim(mtcars)
```

```
## [1] 32 11
```

```
head(mtcars)
```

```
##                    mpg cyl disp  hp drat    wt  qsec
## Mazda RX4         21.0   6  160 110 3.90 2.620 16.46
## Mazda RX4 Wag     21.0   6  160 110 3.90 2.875 17.02
## Datsun 710        22.8   4  108  93 3.85 2.320 18.61
## Hornet 4 Drive    21.4   6  258 110 3.08 3.215 19.44
## Hornet Sportabout 18.7   8  360 175 3.15 3.440 17.02
## Valiant           18.1   6  225 105 2.76 3.460 20.22
##                   vs am gear carb
## Mazda RX4          0  1    4    4
## Mazda RX4 Wag      0  1    4    4
## Datsun 710         1  1    4    1
## Hornet 4 Drive     1  0    3    1
## Hornet Sportabout  0  0    3    2
## Valiant            1  0    3    1
```

We create a new data frame that contains only the numerical variables that we will use. We exclude the 5 categorical variables.

```
DF = subset(mtcars , select=-c(cyl, vs, am, gear,carb))

# view the first six rows of new data frame
head(DF)
```

```
##                    mpg disp  hp drat    wt  qsec
## Mazda RX4         21.0  160 110 3.90 2.620 16.46
## Mazda RX4 Wag     21.0  160 110 3.90 2.875 17.02
## Datsun 710        22.8  108  93 3.85 2.320 18.61
## Hornet 4 Drive    21.4  258 110 3.08 3.215 19.44
## Hornet Sportabout 18.7  360 175 3.15 3.440 17.02
## Valiant           18.1  225 105 2.76 3.460 20.22
```

Another way to eliminate some columns is as follows:

```
# create new data frame that contains only the selected variables:
data =  mtcars[ , c("mpg", "disp", "hp", "drat", "wt", "qsec")]
```

Next, we look for the linear regression that relates the variable "mpg" to the other quantitative variables:

```
Model1= lm(mpg ~ ., data=DF)
Model1
```

```
##
## Call:
## lm(formula = mpg ~ ., data = DF)
##
## Coefficients:
## (Intercept)          disp            hp          drat
##    16.53357       0.00872      -0.02060       2.01577
##          wt          qsec
##    -4.38546       0.64015
```

Testing the Usefulness of Model 1:

Analysis of Variance F-test

Is the regression equation that uses information provided by the predictor variables x_{disp}, x_{hp}, x_{drat}, x_{wt}, x_{qsec} substantially better than the simple predictor $\bar{y}$ that does not rely on any of the x values?

This question is answered using an overall F-test with the hypotheses

$$H_0: \quad \beta_{disp} = \beta_{hp} = \beta_{drat} = \beta_{wt} = \beta_{qsec} = 0 \qquad \text{versus}$$

$$H_a: \quad \begin{array}{l} \text{at least one of the parameters} \\ \beta_{disp} \ \beta_{hp}, \ \beta_{drat}, \ \beta_{wt}, \ \beta_{qsec} \ \text{differs from 0.} \end{array}$$

The value of the test statistic F is found in the summary as:

```
summary(Model1)
```

```
##
## Call:
## lm(formula = mpg ~ ., data = DF)
##
## Residuals:
##     Min     1Q Median     3Q    Max
## -3.540 -1.670 -0.426  1.132  5.500
##
## Coefficients:
##               Estimate Std. Error t value Pr(>|t|)
## (Intercept) 16.53357   10.96423    1.51   0.1436
## disp         0.00872    0.01119    0.78   0.4428
## hp          -0.02060    0.01528   -1.35   0.1894
## drat         2.01577    1.30946    1.54   0.1358
## wt          -4.38546    1.24343   -3.53   0.0016 **
## qsec         0.64015    0.45934    1.39   0.1752
## ---
## Signif. codes:
## 0 '***' 0.001 '**' 0.01 '*' 0.05 '.' 0.1 ' ' 1
##
## Residual standard error: 2.56 on 26 degrees of freedom
## Multiple R-squared:  0.849,  Adjusted R-squared:  0.82
## F-statistic: 29.2 on 5 and 26 DF,  p-value: 6.89e-10
```

The statistic F is

$$F = \frac{MSR}{MSE}$$

has an F distribution with $df_1 = k = 5$ and $df_2 = n - k - 1 = 32 - 5 - 1 = 26$. Its value is: $F = 29.22$. The p-value given in the summary is 6.892×10^{-10}, and can be calculated by:

```
k=5
n=32
F=29.22
pv=pf(F, df1= k, df2=n-k-1, lower.tail=FALSE)
pv
```

```
## [1] 6.886e-10
```

Since the p-value is less than $\alpha = 0.01$, we can declare the regression to be highly significant. That is, at least one of the predictor variables is contributing significant information for the prediction of the response variable y.

For any conventional α level, the null hypothesis is rejected, so it is obvious that if the prediction equation is based on all $k = 5$ regression variables, it is significant.

ANOVA table for the Complete Model 1: To obtain the previous values included in an ANOVA table for this Model 1, one can proceed as follows.

```
Model0 = lm(mpg ~ 1, data = DF)
anova(Model0, Model1)
```

```
## Analysis of Variance Table
##
## Model 1: mpg ~ 1
## Model 2: mpg ~ disp + hp + drat + wt + qsec
##   Res.Df  RSS Df Sum of Sq    F   Pr(>F)
## 1     31 1126
## 2     26  170  5       956 29.2 6.9e-10 ***
## ---
## Signif. codes:
## 0 '***' 0.001 '**' 0.01 '*' 0.05 '.' 0.1 ' ' 1
```

However, this F test does not indicate whether all of the regressor variables are needed for a prediction equation.

Testing the Usefulness of Model 1:

Significance of Coefficients

It is necessary to examine the tests of significance for the individual partial regression coefficients in order to determine their relative importance in explaining the variations of "*mpg*".

The individual t-tests in the summary are designed to test the hypotheses:

$$H_0 : \beta_i = 0 . \qquad \text{versus}$$
$$H_a : \beta_i \neq 0 \qquad i = \ disp, \ hp, \ drat, \ wt, \ qsec$$

for each of the partial regression coefficients, given that the other predictor variables are already in the model. These tests are based on the Student's t statistic given by

$$t = \frac{\widehat{\beta_i} - \beta_i}{\sigma_{\widehat{\beta_i}}}$$

which has $df = n - k - 1$ degrees of freedom.

From the summary, we report the information as follows:

Hypothesis	Coefficient $\hat{\beta}_i$	Standard Error	$T = \hat{\beta}_i / \sigma_{\hat{\beta}_i}$	p−value
$\beta_{disp} = 0$	$\hat{\beta}_{disp} = 0.00872$	$\sigma_{\hat{\beta}_{disp}} = 0.01119$	0.779	0.44281
$\beta_{hp} = 0$	$\hat{\beta}_{hp} = -0.02060$	$\sigma_{\hat{\beta}_{hp}} = 0.01528$	−1.348	0.18936
$\beta_{drat} = 0$	$\hat{\beta}_{drat} = 2.01578$	$\sigma_{\hat{\beta}_{drat}} = 1.30946$	1.539	0.13579
$\beta_{wt} = 0$	$\hat{\beta}_{wt} = -4.38546$	$\sigma_{\hat{\beta}_{wt}} = 1.24343$	−3.527	0.00158
$\beta_{qsec} = 0$	$\hat{\beta}_{qsec} = 0.64015$	$\sigma_{\hat{\beta}_{qsec}} = 0.45934$	1.394	0.17523

From these tests and by examining the p-values, it is seen that only "wt" adds a significant predictive ability to a multiple regression equation which already

contains the variables *disp*, *hp*, *drat*, and *qsec* at no significance level $\alpha = 0.05$;
Only the *p*-value for the test:

$$H_0 : \beta_{wt} = 0 \qquad\qquad \text{versus} \qquad\qquad H_a : \beta_{wt} \neq 0.$$

is less than $\alpha = 0.05$.

Using the Coefficient Plot, we plot confidence intervals for the coefficients β_i
(Figure 2.9). The confidence intervals support the *p*-value approach since only
the confidence interval for β_{wt} doesn't contain zero.

```
knitr::opts_chunk$set(warning = FALSE, message = FALSE)
require(coefplot)
coefplot(Model1 )
```

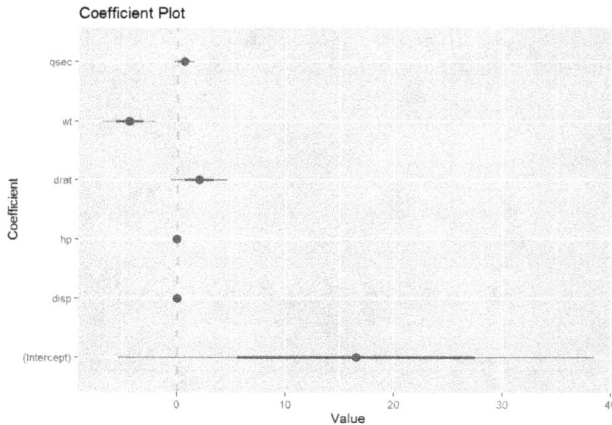

FIGURE 2.9 Confidence Intervals: Model 1 with 5 variables

Testing the Usefulness of Model 1:

R^2 and R^2_{adj}

▶ **The coefficient of determination, R^2.** How well does the regression model fit?

The regression summary provides a statistical measure of the strength of the model in the coefficient of determination R^2, labeled "Multiple R-Squared"—the proportion of the total variation that is explained by the regression of y on x_{disp}, x_{hp}, x_{drat}, x_{wt}, x_{qsec}—defined as

$$R^2 = \frac{SSR}{TotalSS} = 0.8489 \qquad \text{or} \qquad 84.89\%$$

Hence, for this example, 84.89% of the total variation has been explained by the regression model. The model fits well.

▶ **The adjusted value of R^2.**

We have

$$R^2_{adj} = 1 - \left[\frac{n-1}{n-(k+1)}\right]\left(\frac{SSE_C}{S_{yy}}\right) = 1 - \left[\frac{n-1}{n-(k+1)}\right](1-R^2)$$

```
k=5
n=32

R2 = 0.8489
R2adj = 1- ((n-1)/(n-k-1) )*(1-R2)
R2adj
```

```
## [1] 0.8198
```

The value $R^2_{adj} = 81.98\%$ represents the percentage of variation in the response y explained by the independent variable, corrected for degrees of freedom.

Decision for Model 1

We remove the predictor *"disp"* and refit the data. We may consider including the four variables *"hp"*, *"drat"*, *"wt"*, *"qsec"*. Then, we go through the different analyses as above in order to test the usefulness of the new model.

Step 2: A reduced model = Model 2

Equation of Model 2

We fit the data "*mpg*" to the four remaining variables:

```
Model2=lm(mpg ~ hp + drat + wt + qsec , data=DF)
Model2
```

```
##
## Call:
## lm(formula = mpg ~ hp + drat + wt + qsec, data = DF)
##
## Coefficients:
## (Intercept)           hp         drat           wt
##     19.2597      -0.0178       1.6571      -3.7077
##         qsec
##      0.5275
```

Testing the Usefulness of Model 2:

Analysis of Variance F-Test

Is the regression equation that uses information provided by the predictor variables x_{hp}, x_{drat}, x_{wt}, x_{qsec} are substantially better than the simple predictor $\overline{y}$ that does not rely on any of these x values?

This question is answered using an overall F-test with the hypotheses

$$H_0: \quad \beta_{hp} = \beta_{drat} = \beta_{wt} = \beta_{qsec} = 0 \qquad \text{versus}$$

$$H_a: \quad \begin{array}{l} \text{at least one of the parameters} \\ \beta_{hp},\ \beta_{drat},\ \beta_{wt},\ \beta_{qsec}\ \text{differ from 0.} \end{array}$$

The value of the test statistic F is found in the summary as:

```
summary(Model2)
```

```
##
## Call:
## lm(formula = mpg ~ hp + drat + wt + qsec, data = DF)
##
## Residuals:
##    Min     1Q Median     3Q    Max
## -3.578 -1.663 -0.342  1.132  5.442
##
## Coefficients:
##              Estimate Std. Error t value Pr(>|t|)
## (Intercept)  19.2597    10.3154    1.87  0.07279 .
## hp           -0.0178     0.0148   -1.21  0.23732
## drat          1.6571     1.2170    1.36  0.18456
## wt           -3.7077     0.8823   -4.20  0.00026 ***
## qsec          0.5275     0.4328    1.22  0.23347
## ---
## Signif. codes:
## 0 '***' 0.001 '**' 0.01 '*' 0.05 '.' 0.1 ' ' 1
##
## Residual standard error: 2.54 on 27 degrees of freedom
## Multiple R-squared:  0.845,  Adjusted R-squared:  0.822
## F-statistic: 36.9 on 4 and 27 DF,  p-value: 1.41e-10
```

The statistic F is

$$F = \frac{MSR}{MSE}$$

has an F distribution with $df_1 = k = 4$ and $df_2 = n - k - 1 = 32 - 4 - 1 = 27$. Its value is:

$$F = 36.91.$$

The p-value given in the summary is 1.408×10^{-10}, and can be calculated by:

```
k=4
n= 32
F= 36.91
pv=pf(F, df1= k, df2=n-k-1, lower.tail=FALSE)
pv
```

```
## [1] 1.407e-10
```

Since the p-value is less than $\alpha = 0.01$, we can declare the regression to be highly significant. That is, at least one of the predictor variables hp, $drat$, wt, $qsec$ is contributing significant information for the prediction of the response variable mpg.

For any conventional α level, the null hypothesis is rejected, so it is obvious that if the prediction equation is based on all $k = 4$ regression variables, it is significant.

ANOVA table for Model 2: To obtain the previous values included in an ANOVA table for this Model 2, one can proceed as follows.

```
anova(Model0, Model2)
```

```
## Analysis of Variance Table
##
## Model 1: mpg ~ 1
## Model 2: mpg ~ hp + drat + wt + qsec
##   Res.Df  RSS Df Sum of Sq    F  Pr(>F)
## 1     31 1126
## 2     27  174  4       952 36.9 1.4e-10 ***
## ---
## Signif. codes:
## 0 '***' 0.001 '**' 0.01 '*' 0.05 '.' 0.1 ' ' 1
```

Testing the Usefulness of Model 2:

Significance of Coefficients

We examine the tests of significance for the individual partial regression coefficients in order to determine their relative importance in explaining the variations of "*mpg*".

From summary(Model2), we observe in each of the following individual t-test that the null hypothesis is not rejected with p-values greater than 0.05:

$$H_0 : \beta_i = 0 \qquad \text{versus}$$

$$H_a : \beta_i \neq 0 \qquad i = hp, \ drat, \ wt, \ qsec \ .$$

One of the variables may be affecting the others. The variable wt is contributing significant information to Model 2 in the presence of the others. The Confidence Intervals approach β_i supports the p-value approach as shown in the plot (Figure 2.10).

```
require(coefplot)
multiplot(Model1,Model2 )
```

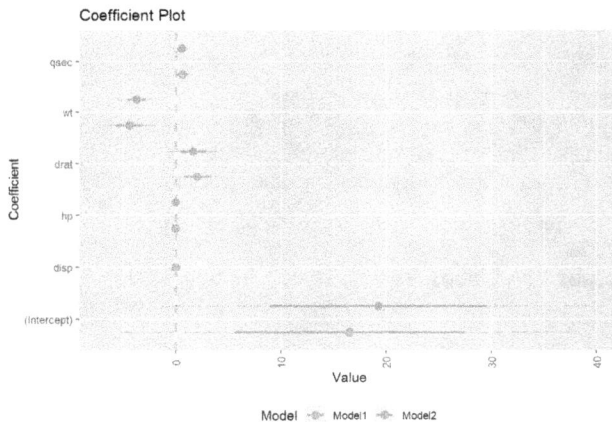

FIGURE 2.10 Confidence Intervals: Model 2 with 4 variables

Testing the Usefulness of Model 2:

R^2 and R^2_{adj}

▶ **The coefficient of determination R^2.**

$$R^2 = \frac{SSR}{TotalSS} = 0.8454 \qquad \text{or} \qquad 84.54\%$$

Notice that R^2 for the full model of 0.8454 is not reduced in model 2. Thus the removal of the predictor "*disp*" doesn't cause a reduction in fit.

▶ **The adjusted value of R^2.** The value $R^2_{adj} = 0.8225$ changed slightly.

```
k=4
n=32
R2= 0.8454
R2adj=1- ((n-1)/(n-k-1) )*(1-R2)
R2adj
```

```
## [1] 0.8225
```

Testing the Usefulness of Model 2:

Comparing Model 1 and Model 2

In order to compare the two models, we test the following hypotheses:

$$H_0: \quad \beta_{disp} = 0 \qquad \text{versus} \qquad H_a: \quad \beta_{disp} \neq 0$$

By comparing the two models, we have the following values:

	Full Model $x_{disp}, x_{hp}, x_{drat}, x_{wt}, x_{qsec}$	Reduced Model $x_{hp}, x_{drat}, x_{wt}, x_{qsec}$
SSE	170.13	174.10
MSE	6.54	6.45
k	5	4
$F = \dfrac{MSR}{MSE}$	29.22	36.91

where k is the number of regressors.

Assume H_0 is true. Then, if SSE_R and SSE_C denote the sum of squares of errors for the reduced model and the complete model, respectively, we have

$$F = \frac{\frac{(SSE_R - SSE_C)/\sigma^2}{k-r}}{\frac{SSE_C/\sigma^2}{n-(k+1)}} = \frac{(SSE_R - SSE_C)/(k-r)}{SSE_C/[n-(k+1)]} \rightsquigarrow \mathcal{F}_{df_1=k-r \; ; \; df_2=n-(k+1)}$$

```
n=32
k=5
r=4

SSE_R= 174.10
SSE_C= 170.13

F=((SSE_R - SSE_C)/(k-r))/(SSE_C/(n-(k+1)) )
F
```

```
## [1] 0.6067
```

The critical value of F with $\alpha = 0.05$ is

```
alpha=0.05
a = qf(alpha, df1=k-r, df2=n-(k+1), lower.tail = FALSE)
a                                      # a=F_{alpha; df1; df2}
```

```
## [1] 4.225
```

```
pf(F, df1=k-r, df2=n-(k+1), lower.tail = FALSE)
```

```
## [1] 0.4431
```

The observed value $F = 0.6067125$ is less than $F_\alpha = 4.225201$. Hence, it does not fall in the rejection region $[F > F_\alpha]$. We conclude that, at the level of significance $\alpha = 0.05$, there is not enough evidence to support a claim that β_{disp} differs from 0.

Hence, we learn that we can explain the variation of the variable *mpg* quite effectively without having to use measurements on the variable *disp*.

Comparing Model 1 and Model 2 using "ANOVA"

```
anova(Model2,Model1)
```

```
## Analysis of Variance Table
##
## Model 1: mpg ~ hp + drat + wt + qsec
## Model 2: mpg ~ disp + hp + drat + wt + qsec
##   Res.Df RSS Df Sum of Sq    F Pr(>F)
## 1     27 174
## 2     26 170  1     3.97 0.61   0.44
```

Next, we remove the variable "hp" with the highest *p*-value of 0.237319 compared to the other variables.

Step 3: A reduced model = Model 3

Equation of Model 3

We fit the data *mpg* to the three remaining variables:

```
Model3=lm(mpg ~  drat + wt + qsec , data=DF)
Model3
```

```
##
## Call:
## lm(formula = mpg ~ drat + wt + qsec, data = DF)
##
## Coefficients:
## (Intercept)          drat            wt          qsec
##      11.394         1.656        -4.398         0.946
```

Testing the Usefulness of Model 3:

Analysis of Variance F-test

Is the regression equation that uses information provided by the predictor variables x_{drat}, x_{wt}, x_{qsec} are substantially better than the simple predictor $\overline{y}$ that does not rely on any of the x values?

This question is answered using an overall F-test with the hypotheses

$$H_0: \quad \beta_{drat} = \beta_{wt} = \beta_{qsec} = 0 \qquad\qquad \text{versus}$$

$H_a:$ at least one of the parameters β_{drat}, β_{wt}, β_{qsec} differs from 0.

The value of the test statistic F is found in the summary as:

```
summary(Model3)
```

```
##
## Call:
## lm(formula = mpg ~ drat + wt + qsec, data = DF)
##
## Residuals:
##    Min    1Q Median    3Q    Max
##  -4.12  -1.83  -0.27  1.05   5.50
##
## Coefficients:
##              Estimate Std. Error t value Pr(>|t|)
## (Intercept)   11.394      8.069    1.41   0.1689
## drat           1.656      1.227    1.35   0.1879
## wt            -4.398      0.678   -6.49    5e-07 ***
## qsec           0.946      0.262    3.62   0.0012 **
## ---
## Signif. codes:
## 0 '***' 0.001 '**' 0.01 '*' 0.05 '.' 0.1 ' ' 1
##
## Residual standard error: 2.56 on 28 degrees of freedom
## Multiple R-squared:  0.837,  Adjusted R-squared:  0.82
## F-statistic: 47.9 on 3 and 28 DF,  p-value: 3.72e-11
```

The statistic F is

$$F = \frac{MSR}{MSE}$$

has an F distribution with $df_1 = k = 3$ and $df_2 = n - k - 1 = 32 - 3 - 1 = 28$. Its value is: $F = 47.93$. The p-value given in the summary is 3.768×10^{-11}, and can be calculated by:

```
k=3
n= 32
F= 47.93
```

```
pv=pf(F, df1= k, df2=n-k-1, lower.tail=FALSE)
pv
```

```
## [1] 3.726e-11
```

Since the p-value is less than $\alpha = 0.05$, we can declare the regression to be highly significant. That is, at least one of the 3 predictor variables is contributing significant information for the prediction of the response variable *mpg*.

For any conventional α level, the null hypothesis is rejected, so it is obvious that if the prediction equation is based on all $k = 3$ regression variables, it is significant.

ANOVA table for Model 3: To obtain the previous values included in an ANOVA table for this Model 3, one can proceed as follows.

```
anova(Model0, Model3)
```

```
## Analysis of Variance Table
##
## Model 1: mpg ~ 1
## Model 2: mpg ~ drat + wt + qsec
##   Res.Df  RSS Df Sum of Sq    F  Pr(>F)
## 1     31 1126
## 2     28  184  3       943 47.9 3.7e-11 ***
## ---
## Signif. codes:
## 0 '***' 0.001 '**' 0.01 '*' 0.05 '.' 0.1 ' ' 1
```

Testing the Usefulness of Model 3:

Significance of coefficients

We examine the tests of significance for the individual partial regression coefficients in order to determine their relative importance in explaining the variations of "*mpg*".

From *summary(Model3)*, we observe in each of the following individual *t*-test that the null hypothesis is rejected with *p*-values less than 0.05:

$$H_0: \quad \beta_i = 0 \qquad \text{versus} \qquad H_a: \quad \beta_i \neq 0 \qquad i = wt, \; qsec.$$

The variable "*drat*" is not contributing well to the regression.

These observations are confirmed by the confidence interval plots for β_i (Figure 2.11).

```
require(coefplot)
multiplot(Model1,Model2,Model3 )
```

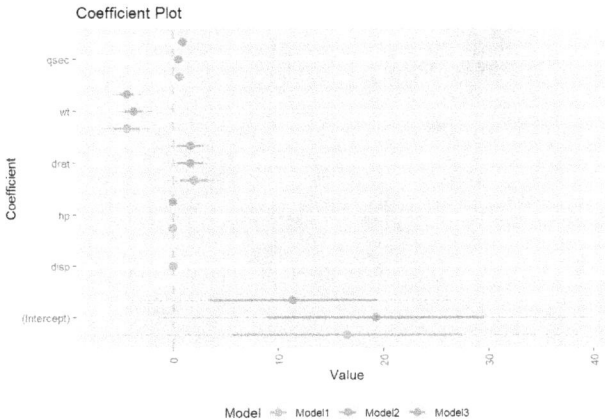

FIGURE 2.11 Confidence Intervals: Model 3 with 3 variables

Testing the Usefulness of Model 3:

R^2 and R^2_{adj}

► **The coefficient of determination, R^2.**

$$R^2 = \frac{SSR}{TotalSS} = 0.837 \qquad \text{or} \qquad 83.7\%$$

Notice that the coefficient of determination R^2 for Model 2 ($R^2_{Model2} = 0.8454$) is slightly reduced in Model 3 ($R^2_{Model3} = 0.837$). Thus the removal of the predictor "hp" doesn't cause an important reduction in fit.

► **The adjusted value of R^2.** The value $R^2_{adj} = 0.8196$ changed slightly.

```
k=3
n=32
R2=0.837
R2adj=1- ((n-1)/(n-k-1) )*(1-R2)
R2adj
```

```
## [1] 0.8195
```

Testing the Usefulness of Model 3:

Comparing Model 2 and Model 3

In order to compare the two models, we test the following null Hypothesis H_0 versus the alternative Hypothesis H_a:

$$H_0: \quad \beta_{hp} = 0 \qquad \text{versus} \qquad H_a: \quad \beta_{hp} \neq 0$$

By comparing the two models, we recall the following values from the summaries of the 3 models:

	Model 1 $x_{disp}, x_{hp}, x_{drat},$ x_{wt}, x_{qsec}	**Model 2** $x_{hp}, x_{drat},$ x_{wt}, x_{qsec}	**Model 3** $x_{drat}, x_{wt},$ x_{qsec}
SSE	170.13	174.10	183.52
MSE	6.54	6.45	6.55
k	5	4	3
$F = \dfrac{MSR}{MSE}$	29.22	36.91	47.93

Assume H_0 is true. Then, if SSE_R and SSE_C denote the sum of squares of errors for the reduced model and the complete model, respectively, we have

$$F = \frac{\frac{(SSE_R - SSE_C)/\sigma^2}{k-r}}{\frac{SSE_C/\sigma^2}{n-(k+1)}} = \frac{(SSE_R - SSE_C)/(k-r)}{SSE_C/[n-(k+1)]} \rightsquigarrow \mathcal{F}_{df_1 = k-r \; ; \; df_2 = n-(k+1)}$$

```
n=32
k=4
r=3

SSE_R=  183.52
SSE_C=  174.10

F=((SSE_R - SSE_C)/(k-r))/(SSE_C/(n-(k+1)) )
F
```

```
## [1] 1.461
```

The critical value of F with $\alpha = 0.05$ is

```
alpha=0.05
a= qf(alpha, df1=k-r, df2=n-(k+1), lower.tail = FALSE)
a                             #a=F_{alpha; df1; df2}
```

```
## [1] 4.21
```

The observed value $F = 1.460885$ is less than $F_\alpha = 4.210008$. Hence, it does not fall in the rejection region $[F > F_\alpha]$. We conclude that, at the level of significance $\alpha = 0.05$, there is not enough evidence to support a claim that β_{hp} differs from 0.

Hence, we learn that we can explain the variation of the variable *mpg* quite effectively without having to use measurements on the variables *disp* and *hp*.

Comparing Model 2 and Model 3 using "ANOVA"

`anova(Model3,Model2)`

```
## Analysis of Variance Table
##
## Model 1: mpg ~ drat + wt + qsec
## Model 2: mpg ~ hp + drat + wt + qsec
##   Res.Df RSS Df Sum of Sq      F Pr(>F)
## 1     28 184
## 2     27 174  1      9.42 1.46   0.24
```

Next, we remove the variable "drat" with the highest *p*-value of 0.198755 compared to the other variables.

Step 4: A reduced model = Model 4

Equation of Model 4

We fit the data *mpg* to the two remaining variables:

```
Model4=lm(mpg ~  wt + qsec , data=DF)
Model4
```

```
##
## Call:
## lm(formula = mpg ~ wt + qsec, data = DF)
##
## Coefficients:
## (Intercept)            wt          qsec
##      19.746        -5.048         0.929
```

Testing the Usefulness of Model 4:

Analysis of Variance F-Test

Is the regression equation that uses information provided by the predictor variables x_{wt}, x_{qsec} are substantially better than the simple predictor $\bar{y}$ that does not rely on any of the x values?

This question is answered using an overall F-test with the hypotheses

$$H_0: \quad \beta_{drat} = 0 \qquad \text{versus} \qquad H_a: \quad \beta_{drat} \neq 0$$

The value of the test statistic F is found in the summary as:

```
summary(Model4)
```

```
##
## Call:
## lm(formula = mpg ~ wt + qsec, data = DF)
##
## Residuals:
##     Min    1Q Median    3Q    Max
## -4.396 -2.143 -0.213  1.492  5.749
##
## Coefficients:
##             Estimate Std. Error t value Pr(>|t|)
## (Intercept)   19.746      5.252    3.76  0.00077 ***
## wt            -5.048      0.484  -10.43  2.5e-11 ***
## qsec           0.929      0.265    3.51  0.00150 **
## ---
## Signif. codes:
## 0 '***' 0.001 '**' 0.01 '*' 0.05 '.' 0.1 ' ' 1
##
## Residual standard error: 2.6 on 29 degrees of freedom
## Multiple R-squared:  0.826,  Adjusted R-squared:  0.814
## F-statistic:  69 on 2 and 29 DF,  p-value: 9.39e-12
```

The statistic F is

$$F = \frac{MSR}{MSE}$$

has an F distribution with $df_1 = k = 2$ and $df_2 = n - k - 1 = 32 - 2 - 1 = 29$. Its value is: $F = 69.03$. The p-value given in the summary is 9.395×10^{-12}, and can be calculated by:

```
k = 2
n = 32
F =   69.03
pv=pf(F, df1= k, df2=n-k-1, lower.tail=FALSE)
pv
```

```
## [1] 9.4e-12
```

Since the p-value is less than $\alpha = 0.01$, we can declare the regression to be highly significant. That is, at least one of the predictor variables is contributing significant information for the prediction of the response variable *mpg*.

For any conventional α level, the null hypothesis is rejected, so it is obvious that if the prediction equation is based on all $k = 2$ regression variables, it is significant.

ANOVA table for Model 4: To obtain the previous values included in an ANOVA table for this Model 4, one can proceed as follows.

```
anova(Model0, Model4)
```

```
## Analysis of Variance Table
##
## Model 1: mpg ~ 1
## Model 2: mpg ~ wt + qsec
##   Res.Df  RSS Df Sum of Sq  F  Pr(>F)
## 1     31 1126
## 2     29  195  2       931 69 9.4e-12 ***
## ---
## Signif. codes:
## 0 '***' 0.001 '**' 0.01 '*' 0.05 '.' 0.1 ' ' 1
```

Testing the Usefulness of Model 4:

Significance of coefficients

We examine the tests of significance for the individual partial regression coefficients in order to determine their relative importance in explaining the variations of "*mpg*".

From the summary, each of the following individual t-test that the null hypothesis is rejected with p-values less than 0.05:

$$H_0: \quad \beta_i = 0 \qquad \text{versus} \qquad H_a: \quad \beta_i \neq 0 \qquad i = wt, \ qsec.$$

Therefore, each of the variables *wt, qsec* contributes significantly to Model 4.

We visualize the models together using multiplot from the coefplot package. The results in Figure 2.12 shows that the confidence intervals for the coefficients β_i in Model4 are far from 0. This confirms that the variables "*wt*" and "*qsec*" have signifiacnt effect on the variable "*mpg*" compared to the other variables.

```
multiplot(Model1, Model2, Model3, Model4, pointSize = 2 )
```

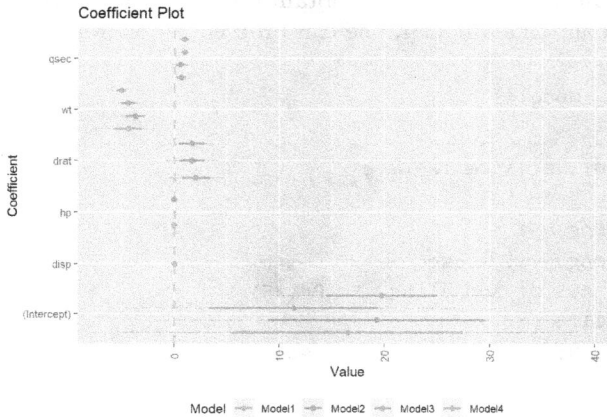

FIGURE 2.12 Confidence Intervals: Model 4 with 2 variables

Testing the Usefulness of Model 4:

$\mathbf{R^2}$ and $\mathbf{R^2_{adj}}$

▶ **The coefficient of determination, R^2.**

$$R^2 = \frac{SSR}{TotalSS} = 0.8268 \qquad \text{or} \qquad 82.68\%$$

Notice that R^2 for Model 4 of $R^2_{Model4} = 0.8268$ is slightly reduced compared to Model 3 ($R^2_{Model3} = 0.8369$). Thus the removal of the predictor "drat" doesn't cause an important reduction in fit.

▶ **The adjusted value of** R^2. The value $R_{adj}^2 = 0.8225$ changed slightly.

```
k=2
n=32
R2=0.8268
R2adj=1- ((n-1)/(n-k-1) )*(1-R2)
R2adj
```

```
## [1] 0.8149
```

Testing the Usefulness of Model 4:

Comparing Model 3 and Model 4

In order to compare the reduced model $mpg = wt + qsec$ to the complete model $mpg = drat + wt + qsec$, we test the hypotheses

$$H_0: \quad \beta_{drat} = 0 \qquad \text{versus} \qquad H_a: \quad \beta_{drat} \neq 0$$

First, we record the following values from the summaries:

	Model 1	Model 2	Model 3	Model 4
	$x_{disp}, x_{hp}, x_{drat},$ x_{wt}, x_{qsec}	$x_{hp}, x_{drat},$ x_{wt}, x_{qsec}	$x_{drat}, x_{wt},$ x_{qsec}	$x_{wt},$ x_{qsec}
SSE	170.13	174.10	183.68	195.05
MSE	6.54	6.45	6.56	6.73
k	5	4	3	2
F	29.22	36.91	47.88	69.21

Assume H_0 is true. Then, if SSE_R and SSE_C denote the sum of squares of errors for the reduced model 4 and the complete model 3, respectively, we have

$$F = \frac{\frac{(SSE_R - SSE_C)/\sigma^2}{k-r}}{\frac{SSE_C/\sigma^2}{n-(k+1)}} = \frac{(SSE_R - SSE_C)/(k-r)}{SSE_C/[n-(k+1)]} \rightsquigarrow \mathcal{F}_{df_1 = k-r \; ; \; df_2 = n-(k+1)}$$

```
n=32
k=3
r=2
```

```
SSE_R =   195.05
SSE_C =   183.68
```

```
F=((SSE_R - SSE_C)/(k-r))/(SSE_C/(n-(k+1)) )
F
```

```
## [1] 1.733
```

The critical value of F with $\alpha = 0.05$ is

```
alpha=0.05
a= qf(alpha, df1=k-r, df2=n-(k+1), lower.tail = FALSE)
a                                    #a=F_{alpha; df1; df2}
```

```
## [1] 4.196
```

The observed value $F = 1.733232$ is less than $F_\alpha = 4.195972$. Hence, it does not fall in the rejection region $[F > F_\alpha]$. We conclude that, at the level of significance $\alpha = 0.05$, there is not enough evidence to support a claim that β_{drat} differs from 0.

Hence, we learn that we can explain the variation of the variable *mpg* quite effectively without having to use measurements on the variables *disp*, *qsec*, and *drat*.

Comparing Model 3 and Model 4 using "anova"

```
anova(Model4, Model3)
```

```
## Analysis of Variance Table
##
## Model 1: mpg ~ wt + qsec
## Model 2: mpg ~ drat + wt + qsec
##   Res.Df RSS Df Sum of Sq   F Pr(>F)
```

```
## 1      29 196
## 2      28 184  1       11.9 1.82   0.19
```

Conclusion

Determining the quality of a model is an important step in the model-building process. Several methods are developed for this purpose. We applied the Least Squares Method (LSM) to fit the data set "mtcars" to a multi-linear model. Then, in order to reduce the number of variables that contributed more to the variation of the variable *mpg*, we applied the backward stepwise method, where we looked at the:

- F test for each model (Analysis of Variance),

- t-tests for the individual partial regression coefficients

- R^2 and R^2_{adj}; the coefficient of determination and its adjusted value.

- F-test of comparison between two models

This set of measures describes different aspects of the model and explains the decisions made.

We found that the best linear model is Model 4:

$$mpg \approx 19.7462 - 5.0480\ wt - 0.9292\ qsec.$$

The comparison between steps can be summarized in an ANOVA table:

```
anova(Model0, Model4, Model3, Model2, Model1)
```

```
## Analysis of Variance Table
##
## Model 1: mpg ~ 1
## Model 2: mpg ~ wt + qsec
## Model 3: mpg ~ drat + wt + qsec
## Model 4: mpg ~ hp + drat + wt + qsec
```

```
## Model 5: mpg ~ disp + hp + drat + wt + qsec
##   Res.Df  RSS Df Sum of Sq     F  Pr(>F)
## 1      31 1126
## 2      29  195  2       931 71.11 2.9e-11 ***
## 3      28  184  1        12  1.83    0.19
## 4      27  174  1         9  1.44    0.24
## 5      26  170  1         4  0.61    0.44
## ---
## Signif. codes:
## 0 '***' 0.001 '**' 0.01 '*' 0.05 '.' 0.1 ' ' 1
```

We notice an increase of residuals of the sum of squares as the number of variables decreases in the model while the F test improves.

There exist other metrics for comparing models like the Akaike Information Criterion (AIC) and the Bayesian Information Criterion (BIC).

The "backward-step" function leads to $mpg \approx drat + wt + qsec$ as the best model where the AIC criterion is used.

```
backward=step(Model1,direction='backward',scope=formula(Model1))
```

```
## Start:  AIC=65.47
## mpg ~ disp + hp + drat + wt + qsec
##
##          Df Sum of Sq RSS  AIC
## - disp   1       4.0 174 64.2
## <none>              170 65.5
## - hp     1      11.9 182 65.6
## - qsec   1      12.7 183 65.8
## - drat   1      15.5 186 66.3
## - wt     1      81.4 252 76.0
##
## Step:  AIC=64.21
## mpg ~ hp + drat + wt + qsec
##
##          Df Sum of Sq RSS  AIC
## - hp     1       9.4 184 63.9
## - qsec   1       9.6 184 63.9
## <none>              174 64.2
## - drat   1      12.0 186 64.3
## - wt     1     113.9 288 78.3
##
## Step:  AIC=63.89
## mpg ~ drat + wt + qsec
##
```

```
##           Df Sum of Sq RSS  AIC
## <none>                184 63.9
## - drat  1       11.9 195 63.9
## - qsec  1       85.7 269 74.2
## - wt    1      275.7 459 91.2
```

backward

```
##
## Call:
## lm(formula = mpg ~ drat + wt + qsec, data = DF)
##
## Coefficients:
## (Intercept)         drat          wt        qsec
##      11.394        1.656      -4.398       0.946
```

3

Non-Basic Regression

In this part, we illustrate two methods involving indirectly multilinear regressions. For further exploration, we refer the reader to Panik (2010) for more methods and A. Gelman (2006) for multilevel models in particular.

3.1   Nonlinear Models

Linearizable Nonlinear Model

There are several curvilinear models that describe well some data sets. Each of the following models can be reduced to a linear model by a particular transformation as shown below.

	Nonlinear Model	Transformation to a Linear Model
Power model	$y = ax^b$	$\ln(y) = \ln(a) + b\ln(x)$
Exponential model	$y = ae^{bx}$	$\ln(y) = \ln(a) + bx$
Logistic model	$y = \dfrac{L}{1 + e^{a+bx}}$	$\ln\left(\dfrac{L-y}{y}\right) = a + bx$

A justification for the use of a transformation θ comes from the mean value theorem:

$$\theta(f(x_i))-(mx_i+p) = \theta(f(x_i))-\theta[\theta^{-1}(mx_i+p)] = \theta'(c_i)\left[f(x_i)-\theta^{-1}(mx_i+p)\right].$$

This allows us to compare the two sums of squares

$$\sum\left(\theta(f(x_i)) - (mx_i + p)\right)^2 = \sum(\theta'(c_i))^2\left[f(x_i) - \theta^{-1}(mx_i + p)\right]^2,$$

knowing that (m,p) is a minimum point for the transformed function $\theta \circ f$. Some specific properties of the function θ are needed for that K. Sydsæter (2004).

Many more linearizable nonlinear models are possible, like the **Ratio Model**

$$y = \frac{ax}{b+x}$$

that involves the transformation

$$\frac{1}{y} = \frac{b+x}{ax} = \frac{1}{a} + \left(\frac{b}{a}\right)\left(\frac{1}{x}\right).$$

Non-Linearizable Non-linear Model

Some models are not linearizable by a transformation. The methodology for finding a Nonlinear regression method follows the same principles of maximum likelihood principles. If the errors are normally distributed, the best point estimator of the parameter $\theta = (\theta_1, \ldots, \theta_m)$ is the minimizer of the sum of squares of the residuals:

$$SSE = \sum_{i=1}^{n}[y_i - f(x_i, \theta)]^2.$$

This leads to solving the complicated nonlinear system $\nabla_\theta SSE = 0$.

Using the approximation for f in SSE turns approximately the minimization problem into a linear one. Numerical methods are involved using the Taylor linearization formula of f at a particular value θ_0:

$$f(x, \theta) \approx f(x, \theta_0) + \sum_{k=1}^{m} \frac{\partial f}{\partial \theta_k}(x, \theta_0)(\theta - \theta_0).$$

The initial problem is then transformed by Taylor approximation into a linear or multilinear regression model. The process is iterative, in which the minimizer of the approximate SSE becomes the new estimate of θ. The iterative process is continued until certain stopping criteria are satisfied. The statistical inferences are based on the linear approximation S. Chatterjee (2013).

In what follows, we illustrate some one-variable linearizable nonlinear models formulated from R.J. Larsen (2001).

TABLE 3.1 University tuitions

year	year after 1981, x	tuition in thousands dollars, y
1982	1	6.10
1983	2	6.80
1984	3	7.50
1985	4	8.50
1986	5	9.30
1987	6	10.50
1988	7	11.50
1989	8	12.62
1990	9	13.97
1991	10	14.97

Solved Problems

Case Study: University Tuition

The following are the tuition that were charged at Vanderbilt University from 1982 to 1991 (see Table 3.1).

i) Plot the data.

ii) To fit these data with a model of the form $y = \beta_0 e^{\beta_1 x}$, find the least squares line of best fit of $\ln(y)$ as a function of $\ln(x)$. Deduce approximate values of β_0 and β_1.

iii) Superimpose the answer to part (ii) on the graph to part (i).

iv) Suppose a freshman entered Vanderbilt in 1991, graduated four years later, got married in 3 years, and had a daughter 2 years after that. Based on an extrapolation of the data from 1982 to 1991, what might he expect his daughter's tuition bill to be for 4 years at Vanderbilt? Assume that she enrolls at age 18. Predict the tuition fees amount for his daughter with a 95% confidence interval.

(This question is formulated from R.J. Larsen (2001).)

Solution. i) The plot of the data shows that the tuition fees increase over years (see Figure 3.1).

```
x=c(1, 2, 3, 4, 5, 6, 7, 8, 9, 10)
y=c(6.1, 6.8, 7.5, 8.5, 9.3, 10.5, 11.5, 12.625, 13.975, 14.975)
plot(x,y,  col = "blue", xlab="x: year after 1981",
 ylab="y: tuition (in thousands $)", main="University tuitions",
                        col.main="darkgreen")
```

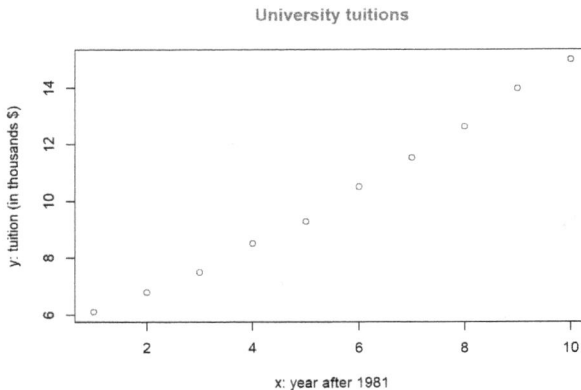

FIGURE 3.1 Tuitions versus Years

ii) We have

$$y = \beta_0 e^{\beta_1 x} \qquad \Longleftrightarrow \qquad \ln(y) = \ln(\beta_0) + \beta_1 x$$

To identify β_0 and β_1, we look for the best line that fits the data $(x, \ln(y))$ and obtain:

```
z= log(y)

sxz=sum((x-mean(x))*(z-mean(z)))
sxx=sum((x-mean(x))*(x-mean(x)))
szz=sum((z-mean(z))*(z-mean(z)))
```

```
b = sxz/sxx
a = mean(z)- b*mean(x)
cat("b=", b, "     ", "a =", a , "      ", "exp(a) =", exp(a))
```

```
## b= 0.1016        a = 1.72        exp(a) = 5.583
```

$$b = \beta_1 = 0.1015597 \quad \text{and}$$
$$\left(a = \ln(\beta_0) = 1.719746 \quad \Longleftrightarrow \quad \beta_0 = e^a \approx 5.583112 \right).$$

The following graph shows how the points are very close to the line (see Figure 3.2).

```
plot(x,z,  col = "blue", xlab="x: year after 1981",
     ylab="z: log(tuition (in thousands $))",
     main="Log(Tuitions) versus Years", col.main="darkgreen")
curve(1.719746 + 0.1015597*x, from=1, to=10, add=TRUE,
      col="red",lwd=2)
legend('bottomright',inset=0.05,c("z=log(y)=1.719746 + 0.1015597 x"),
      lty=1,col=c("red"),title="best line fit")
```

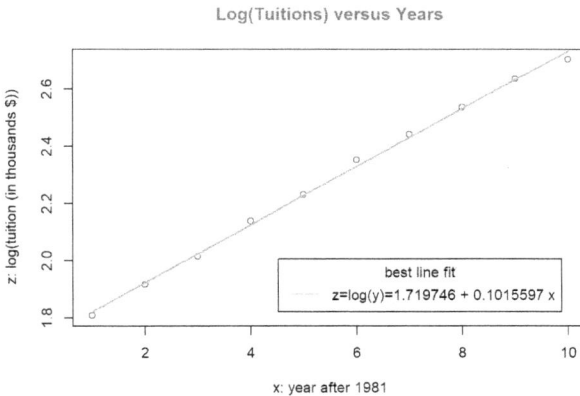

FIGURE 3.2 Log(Tuitions) versus Years

iii) In Figure 3.3, we plot the original data with the exponential function model

```
plot(x,y, col = "blue", xlab="x: year after 1981",
    ylab="y: tuition (in thousands $)",
    main="Tuitions versus Years", col.main="darkgreen")
curve(5.583112*exp(0.1015597*x), 0,100, add=TRUE, col = "red", lwd=2)
legend('bottomright',inset=0.05,c("y=5.583112 e^(0.1015597x)"),
       lty=1,col=c("red"),title="best curve fit")
```

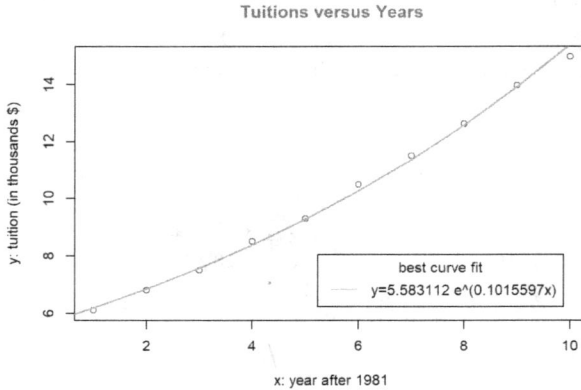

FIGURE 3.3 Tuitions versus Years: Exponential Model

iv) Using the exponential model, the daughter's tuition bill for 4 years at Vanderbilt will be as follows:

After 27 years, the freshman's daughter will be 18 years old and will enter university on the year 2018. The tuition fees after 4 years at university will be 117511.6\$ using the exponential model:

```
x0= 30
y0=5.583112*exp(0.1015597*x0)
y0
```

```
## [1] 117.5
```

The 95% confidence interval, for the predicted value y_0, is obtained as below:

```
n=length(x)
SSR=sxz*sxz/sxx
SSE=szz-SSR
MSE=SSE/(n-2)

var=MSE*(1 + 1/n + (x0-mean(x))*(x0-mean(x))/sxx)
s0=sqrt(var)

t0=qt(p=0.05/2, df=n-2, lower.tail = FALSE, log.p = FALSE)

LCB= log(y0) - (t0)*s0           #  Lower bound of the C.I
UCB= log(y0) + (t0)*s0           #  Upper bound of the C.I
cat("LCB =", LCB, "       ", "UCB=", UCB)

## LCB = 4.661        UCB= 4.872

cat("exp(LCB) =", exp(LCB), "      ", "exp(UCB)=", exp(UCB))

## exp(LCB) = 105.8        exp(UCB)= 130.6
```

The *C.I.* is $(4.661144, 4.871931)$. That is $4.661144 < \log(y) < 4.871931$ or

$$105.757 < y < 130.5728 \quad \text{in thousands of \$.}$$

Although we do not know whether the particular interval $(105.757, 130.5728)$ contains the true prediction y, the procedure that generates it yields intervals that do capture the predicted value.

In the long run, 95% of the intervals constructed in this way will contain the true y when $x = 30$.

TABLE 3.2 Melanoma rates in USA areas

Location number	x	y
1	32.8	9.0
2	33.9	5.9
3	34.1	6.6
4	37.9	5.8
5	40.0	5.5
6	40.8	3.0
7	41.7	3.4
8	42.2	3.1
9	45.0	3.8

Case Study: Skin Cancer

One of the factors thought to contribute to the incidence of skin cancer is ultraviolet (UV) radiation coming from the sun. It is well known that the amount of UV radiation a person receives is a function of the shielding thickness of the earth's ozone layer, which, in turn, depends on the person's latitude. Listed below are the malignant skin cancer (melanoma) rates (per 100 000) (y) for white males determined for nine areas throughout the United States during the three-year period from 1969 to 1971. The locations of each area are given in "degrees north latitude" (x) (see Table 3.2).

i)   Use the table to plot the points (x, y). Do the points appear to lie along a straight line?

The exponential model, $y = ae^{bx}$, or equivalently $\ln(y) = \ln(a) + bx$, describes the situation well.

ii)   Complete the table with $z = \ln y$.

iii)   Plot the points (x, z). Do the points appear to lie along a straight line?

iv)   Find the least squares line of best fit of z as a function of x.

v)   Plot the points and the regression line on the same graph. Is there sufficient evidence to indicate that there is a correlation between these two variables? Test at the 5% level of significance.

vi)   Find the exponential model function $y = f(x) = ae^{bx}$.

vii)   Plot the points (x, y) from the table and the graph of the function f on the same screen.

(This problem is formulated from R.J. Larsen (2001) Question 10.4.10.)

Solution.

i) *Plotting the data:* (see Figure 3.4).

```
x=c(32.8, 33.9, 34.1, 37.9, 40.0, 40.8, 41.7, 42.2, 45.0)
y=c(9.0, 5.9, 6.6, 5.8, 5.5, 3.0, 3.4, 3.1, 3.8)
plot(x,y,  col = "blue", xlab="x: Latitude", ylab="y: Melanoma rate",
        lwd = 2, main="Skin cancer and UV", col.main="brown")
```

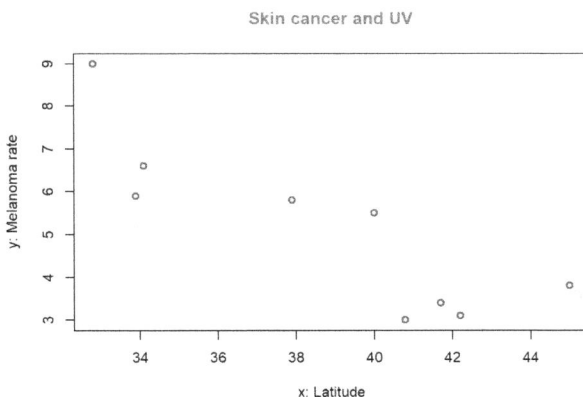

FIGURE 3.4 Skin cancer and UV

The points show a certain pattern, but not necessarily a pattern with a linear tendency.

ii), iii) and **vi)** We have

$$y = ae^{bx} \qquad \Longleftrightarrow \qquad \ln(y) = \ln(a) + bx$$

To identify a and b, we look for the best line that fits the data $(x, \ln(y))$ and obtain:

```
z=log(y)
sxz=sum((x-mean(x))*(z-mean(z)))
sxx=sum((x-mean(x))*(x-mean(x)))
szz=sum((z-mean(z))*(z-mean(z)))
b = sxz/sxx                  # slope
t = mean(z)- b*mean(x)       # intercept  = ln(a)
cat("b=", b, "       ", "t =", t, "     ", "a=exp(t)=", exp(t))
```

```
## b= -0.07609      t = 4.513        a=exp(t)= 91.24
```

v) To determine whether there is a significant relationship between the variables x and $\ln(y)$, we test the hypothesis

$$H_0 : \rho = 0 \qquad \text{versus} \qquad H_1 : \rho \neq 0$$

The Critical Value Approach: The suitable statistic for the test is

$$T = \sqrt{\frac{(n-2)R^2}{1-R^2}}, \qquad T \quad \text{is} \quad T_{n-2}$$

The Rejection region C, at the $\alpha = 5\%$ level of significance, is characterized by

$$\alpha = P(|T| > t_{\alpha/2}) \quad \text{and}$$
$$C = \{z = \ln(y) : t < -t_{\alpha/2}\} \cup \{z = \ln(y) : t > t_{\alpha/2}\}$$

- The t-value $t_0 = t_{\alpha/2}$ is equal to:

```
n= length(x)
t0=qt(p=0.05/2, df=n-2, lower.tail = FALSE, log.p = FALSE)
t0
```

```
## [1] 2.365
```

- If H_0 was true, then the observed statistic is

```
R=sxz/sqrt(sxx*szz)
t=R*sqrt((n-2)/(1-R*R))
t
```

```
## [1] -4.247
```

- Decision. The observed value $t = -4.246661 < -2.364624 = -t_{\alpha/2}$. It falls in the rejection region and H_0 is rejected.

The data presents sufficient evidence to indicate that a strong relationship exists.

Correlation coefficient:

```
R2=sxz*sxz/(sxx*szz)
R2
```

```
## [1] 0.7204
```

We have $R^2 = 72.03\%$; that is, 72% of the total sum of squares of deviations was reduced by using the least squares equation instead of $\bar{z}$ as a predictor of z.

The p-value approach: We have

$$p_{value} = 2 \times P(T > t) = 2 \times (\text{area to the right of the observed value } t).$$

```
pv=2*pt( q=t, df=n-2, lower.tail = TRUE)
pv
```

```
## [1] 0.003809
```

H_0 is rejected at the level $p < 0.01$. The correlation is declared.

The graph below shows the points very close to the line (see Figure 3.5).

```
plot(x,z, col = "blue", xlab="x: Latitude",
    ylab="z: log(Melanoma rate)", lwd = 2,
    main="UV and Skin cancer", col.main="brown")
curve(4.513492 -0.07609294*x,from = 30,to = 50,
                add=TRUE, col="green", lwd = 2)
legend('bottomleft',inset=0.05,c("z = 4.513492 - 0.07609294 x"),
        lty=1,col=c("green"),title="best line fit")
```

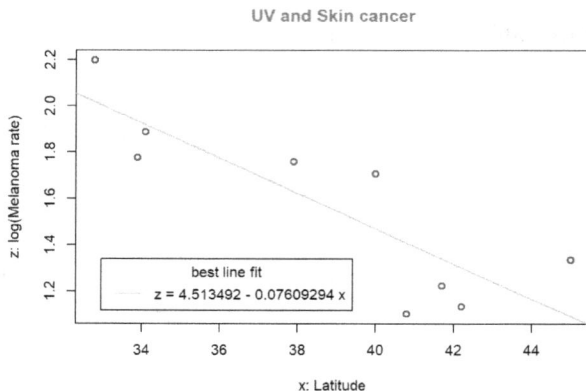

FIGURE 3.5 log(Melanoma rates) and Latitude

The graph shows that the more the thickness of the earth's ozone layer is, the more skin cancer's rate is important.

vi), vii) In this graph, we plot the original data with the exponential model: $y = 91.23984e^{-0.07609294\,x}$ (see Figure 3.6).

```
plot(x,y, col = "blue", xlab="x: Latitude", ylab="y: Melanoma rate",
            lwd = 2, main="UV and Skin cancer", col.main="brown")
curve(91.23984*exp(-0.07609294*x), from = 30, to = 50, add=TRUE,
    col = "green", lwd = 2)
legend('topright',inset=0.05,c("y = 91.23984*exp(-0.07609294 x)"),
            lty=1,col=c("green"),title="best curve fit")
```

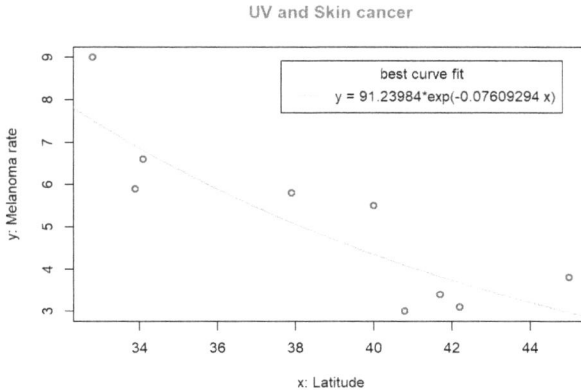

FIGURE 3.6 Melanoma rates and Latitude

TABLE 3.3 Missile force growth

Year, x	Number of ICBM's, y
1960	18
1961	63
1962	294
1963	424
1964	834
1965	854
1966	904
1967	1054
1968	1054
1969	1054

Case Study: Missile force growth

The growth of the American international ballistic missile force during the 1960s is summarized in Table 3.3).

i)   Use the table to plot the points (x, y). Do the points appear to lie along a straight line?

The logistic model, $y = \dfrac{L}{1 + e^{a+bx}}$, describes well the situation.

ii)   With $L = 1060$, use a suitable transformation to express $z = a + bx$ with respect to y.

iii)   Plot the points (x, z). Do the points appear to lie along a straight line?

iv)   Find the least squares line of best fit of z as a function of x.

v)   Plot the points and the regression line on the same graph.

vi)   Find the Logistic model function $y = f(x)$.

vii)   Plot the points (x, y) from the table and the graph of the function f on the same screen.

viii)   Construct the ANOVA table for the linear regression.

ix)   Compare the correlation coefficients of the linear regression for the points (x, y) and (x, z).

(This problem is formulated from R.J. Larsen (2001) Question 10.4.13.)

Solution.

i) *Plotting the data:* (see Figure 3.7).

```
x=c(1960, 1961, 1962, 1963, 1964, 1965, 1966, 1967, 1968, 1969)
y=c(18, 63, 294, 424, 834, 854, 904, 1054, 1054, 1054)
plot(x,y, col = "blue", xlab="x: Year", ylab="y: Number of ICBM's",
     lwd = 2,main="USA Missile force growth", col.main="darkgreen")
```

ii) We have

$$y = \frac{L}{1 + e^{a+bx}} \qquad \Longleftrightarrow \qquad z = \ln\left(\frac{L-y}{y}\right) = a + bx$$

USA Missile force growth

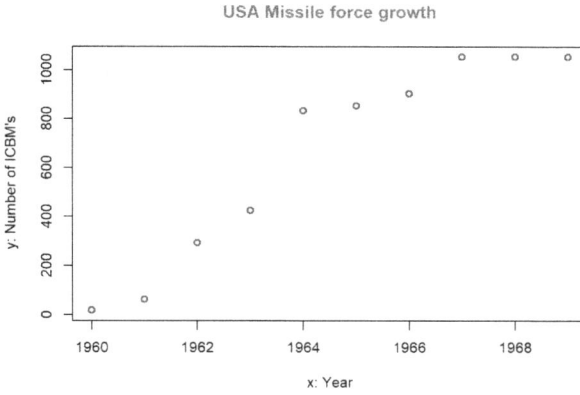

FIGURE 3.7 USA Missile force growth

iii) The plot of the points (x, z) shows a negative linear relationship tendency (see Figure 3.8).

```
L=1060
z=log((L-y)/y)
plot(x,z, col="blue", xlab="x: Year",
         ylab="z: Log((L-y)/y)",
    lwd = 2, main="USA Missile force growth", col.main="darkgreen")
```

USA Missile force growth

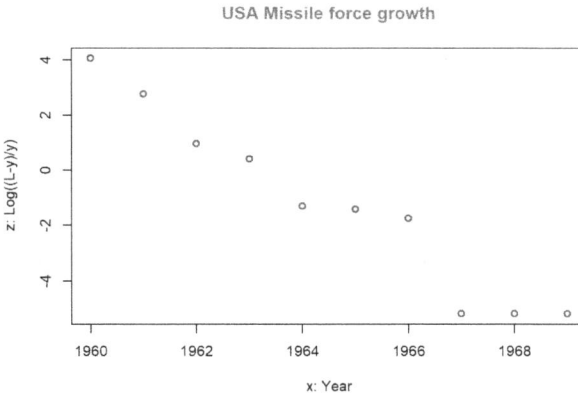

FIGURE 3.8 log((L-growth)/growth) versus years

iv) To identify the coefficients a and b, we look for the best line that fits the data $(x, \ln(\frac{L-y}{y}))$ and obtain:

```
sxz=sum((x-mean(x))*(z-mean(z)))
sxx=sum((x-mean(x))*(x-mean(x)))
szz=sum((z-mean(z))*(z-mean(z)))
b = sxz/sxx                      # slope
a = mean(z)- b*mean(x)           # intercept
cat("b = ", b, "          ", "a = ", a)
```

```
## b =  -1.065           a =  2092
```

Thus the equation of the line and the corresponding logistic model are respectively:

$$\ln(\frac{1060-y}{y}) = 2091.787 - 1.065394\,x \qquad \text{and}$$

$$y = \frac{1060}{1 + e^{2091.787 - 1.065394\,x}}.$$

v) The graph below shows the points very close to the line (see Figure 3.9).

```
plot(x,z, col="blue", xlab="x: Year", ylab="z: Log((L-y)/y)",
       lwd = 2, main="USA Missile force growth", col.main="darkgreen")
curve(2091.787-1.065394*x, 1960,1970, add=TRUE, col="red", lwd=2)
legend('bottomleft',inset=0.05,c("z = 2091.787  - 1.065394 x"),
                lty=1,col=c("red"),title="best line fit")
```

vi), vii) In this graph (see Figure 3.10), we plot the original data with the Logistic function:

```
plot(x,y, col="blue", xlab="x: Year", ylab="y: Number of ICBM's",
     lwd=2,main="USA Missile force growth", col.main="darkgreen")
curve(1060/(1+exp(2091.787-1.065394*x)), 1950,1970, add=TRUE,
      col="red", lwd=2)
legend('bottomright',inset=0.008,
```

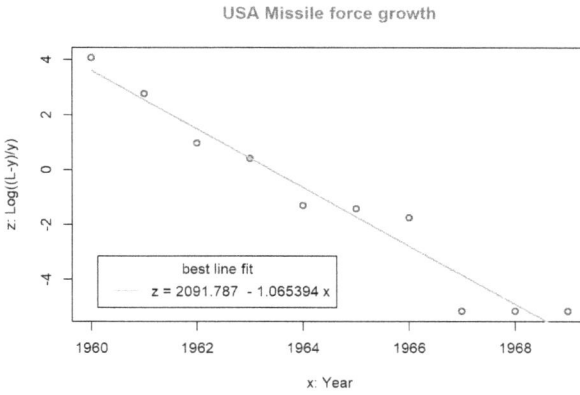

FIGURE 3.9 log((L-growth)/growth) versus years: linear model

```
c("y =1060/(1+exp(2091.787-1.065394*x))"),
lty=1,col=c("red"),title="logistic fit")
```

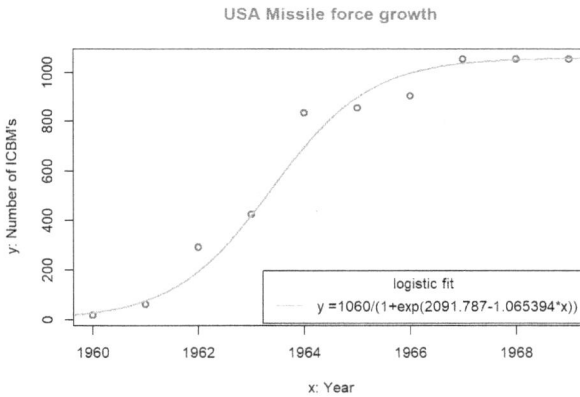

FIGURE 3.10 USA Missile force growth: logistic model

The points are closer to the curve.

viii) The ANOVA table for the (x, z) data set is:

```
n=length(x)

SSR=sxz*sxz/sxx
MSR=SSR
SSE=szz-SSR
MSE=SSE/(n-2)

#ANOVA Table
V1=c("Regression", "Error")
V2=c(1,n-2)
V3=c(SSR, SSE)
V4=c(MSR,MSE)

ANOVA=data.frame(Source=V1, df=V2,  SS=V3,  MS=V4 )
ANOVA


##          Source df    SS       MS
## 1 Regression  1 93.643 93.6429
## 2        Error  8  4.552  0.5691
```

xi) • The correlation coefficient for the (x, z) data set is given by:

```
R2_logistic=sxz*sxz/(sxx*szz)
R2_logistic


## [1] 0.9536
```

that is, 95.36% of the total sum of squares of deviations was reduced by using the least squares equation instead of $\bar{z}$ as a predictor of z.

• The correlation coefficient for the (x, y) data set is:

```
sxy=sum((x-mean(x))*(y-mean(y)))
sxx=sum((x-mean(x))*(x-mean(x)))
syy=sum((y-mean(y))*(y-mean(y)))

R2_linear=sxy*sxy/(sxx*syy)
R2_linear
```

```
## [1] 0.9048
```

That is, 90.48% of the total sum of squares of deviations was reduced by using the least squares equation instead of $\overline{y}$ as a predictor of y.

By comparing the two coefficients, we see that the Logistic model gives a better approximation. Indeed, we have

$$R^2_{Linear\,Model} = 90.48\% \; < \; 95.36\% = R^2_{Logistic\,Model}.$$

3.2   Quantitative and Qualitative Predictors in a Regression Model

Multiple regression has the flexibility to describe both qualitative and quantitative predictor variables.

The tool for that is the use of an "Indicator" (William Mendenhall, 2018) or a "Dummy" (James T. McClave, 2013) variable to represent the categorical variable.

• For example, if a dataset explores the relationship between a numerical variable y with one group of numerical variable A and three groups of categorical variables A_1, A_2, A_3, then one can assign a variable x for group A and indicator variables x_1 and x_2 to describe the groups A_1 and A_2, respectively. The indicator variables x_1 and x_2 are defined by:

$$x_1 = \begin{cases} 1 & \text{if group } A_1 \\ 0 & \text{if not in } A_2 \text{ or } A_3 \end{cases} \qquad x_2 = \begin{cases} 1 & \text{if group } A_2 \\ 0 & \text{if not in } A_1 \text{ or } A_3 \end{cases}$$

Then the linear model describing the data set is:

$$E(y) = \beta_0 + \beta x + \beta_1 x_1 + \beta_2 x_2.$$

For group A_1, the relation will be: $E(y) = \beta_0 + \beta x + \beta_1(1) + \beta_2(0)$.

For group A_2, the relation will be: $E(y) = \beta_0 + \beta x + \beta_1(0) + \beta_2(1)$.

For group A_3, the relation will be: $E(y) = \beta_0 + \beta x + \beta_1(0) + \beta_2(0)$.

If we want to explore the interaction between the variables x_1 and x_2, for example, we can include in the model the interaction term $x_1 x_2$. The model would be:

$$E(y) = \beta_0 + \beta x + \beta_1 x_1 + \beta_2 x_2 + \beta_{12} x_1 x_2$$

which allows the relationship between y and x_1 to behave differently for the group A_2.

• When a quantitative variable involves k categories or levels, $(k-1)$ indicator variables should be added to the regression model.

• Using a single model representing different responses allows testing whether the curves at each level of the categorical variables are different (S. Chatterjee, 2013).

Now, we illustrate this method through the following questions.

TABLE 3.4 Meat consumption

year	Beef	Chicken
1	85	37
2	89	36
3	76	47
4	76	47
5	68	62
6	67	74
7	60	79

Solved Problems

Less Red Meat

Researchers tracked beef and chicken consumption over a period of 7 years. A summary of their data is shown in Table 3.4.

i)   Plot the data in the same window. Compare the consumption of beef and chicken.

ii)   Fit the data to the model

$$E(y) = \beta_0 + \beta_1 x_1 + \beta_2 x_2 + \beta_3 x_1 x_2$$

where y is the annual meat (either beef or chicken in pounds) consumption per person,

$$x_1 = \begin{cases} 1 & \text{if beef} \\ 0 & \text{if chicken} \end{cases} \qquad \text{and} \qquad x_2 = \text{Year}.$$

Use relevant statistics and diagnostic tools to see how well the model fits.

iii)   Write the equations of the two straight lines. On the same graph, plot the data and the least squares prediction equation associated with level 1 (beef) and level 2 (chicken).

TABLE 3.5 Meat Consumption: indicator variable

year	Consumption	Type
1	85	beef
2	89	beef
3	76	beef
4	76	beef
5	68	beef
6	67	beef
7	60	beef
1	37	chicken
2	36	chicken
3	47	chicken
4	47	chicken
5	62	chicken
6	74	chicken
7	79	chicken

iv)   Use the prediction equation to find a point estimate of the average per-person beef consumption in year 8. Find a 95% confidence interval and a 95% prediction interval for that point.

(This question is formulated from William Mendenhall (2018).)

Solution. First, we transform the table to the following one using the indicator variable x_1.

```
y=c(85, 89, 76, 76, 68, 67, 60, 37, 36, 47, 47, 62, 74, 79)
x1type=c("beef","beef", "beef", "beef","beef","beef","beef",
        "chicken","chicken","chicken","chicken","chicken",
        "chicken","chicken")
x2=c(1, 2, 3, 4, 5, 6, 7, 1, 2, 3, 4, 5, 6, 7)
M=data.frame(x2, y, x1type)
colnames(M)=c("year" , "Consumption" , "Type")
```

i)   Figure 3.11 shows the plot of the data in one window. The two categories of meat consumption behave differently. The chicken consumption increases while the beef consumption decreases. Both categories have a linear tendency relationship as years increase.

```
plot(M$year, M$Consumption, main="Meat Comsumption",
    col.main="darkgreen", pch=19, col=as.numeric(factor(x1type)))
legend(x=5,y=50, legend=levels(factor(x1type)),
        text.col=as.numeric(factor(levels(factor(x1type))) ))
```

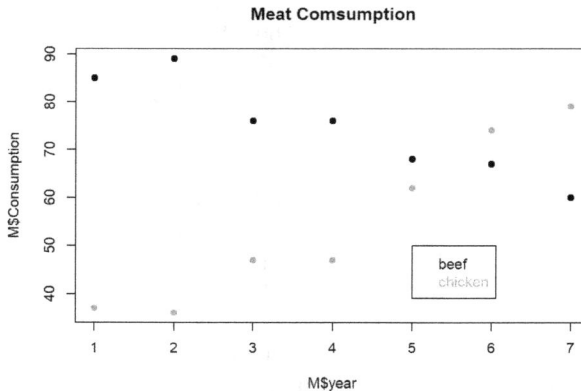

FIGURE 3.11 Meat Consumption

ii) To fit the data to the model, we follow the steps below.

Reading the data

```
x1=c(1, 1, 1, 1, 1, 1, 1, 0, 0, 0, 0, 0, 0, 0)  # Indicator variable
x1x2 = x1*x2
DF=data.frame(x1,x2, x1x2)
```

Analysis of Variance

```
Model0 = lm(y ~ 1, DF)
Model1 = lm(y ~ x1 + x2 + x1x2, DF)
anova(Model0, Model1)

## Analysis of Variance Table
##
## Model 1: y ~ 1
## Model 2: y ~ x1 + x2 + x1x2
```

```
##    Res.Df  RSS Df Sum of Sq    F  Pr(>F)
## 1      13 3812
## 2      10  174  3      3638 69.8 5.2e-07 ***
## ---
## Signif. codes:
## 0 '***' 0.001 '**' 0.01 '*' 0.05 '.' 0.1 ' ' 1
```

Coefficients and their significance

```
summary(Model1)
```

```
##
## Call:
## lm(formula = y ~ x1 + x2 + x1x2, data = DF)
##
## Residuals:
##    Min     1Q Median     3Q    Max
## -7.571 -2.696 -0.071  1.625  5.679
##
## Coefficients:
##             Estimate Std. Error t value Pr(>|t|)
## (Intercept)   23.571      3.522    6.69 5.4e-05 ***
## x1            69.000      4.981   13.85 7.5e-08 ***
## x2             7.750      0.787    9.84 1.8e-06 ***
## x1x2         -12.286      1.114  -11.03 6.4e-07 ***
## ---
## Signif. codes:
## 0 '***' 0.001 '**' 0.01 '*' 0.05 '.' 0.1 ' ' 1
##
## Residual standard error: 4.17 on 10 degrees of freedom
## Multiple R-squared:  0.954,  Adjusted R-squared:  0.941
## F-statistic: 69.8 on 3 and 10 DF,  p-value: 5.21e-07
```

Regression Equation

The second-order model is given by:

$$E(y) = 23.5714 + 69.0000x_1 + 7.7500x_2 - 12.2857x_1x_2$$

Interpretation of the regression coefficients

All the variables are statistically significant with p-values less than 0.05. In particular, the association x_1x_2 detects a difference in the relationships of y within the categories over time. The coefficient of determination R^2 shows that 95.44% of the total variation is explained by the regression model. The model fits well the data.

iii)

• Substituting $x_1 = 0$ into the equation, we obtain the mean consumption of Chicken line:

$$E(y) = 23.5714 + 7.7500x_2.$$

Each one year increase is associated with an increase of 7.75 pounds (per person) of chicken consumption.

• Substituting $x_1 = 1$ into the equation, we obtain the mean consumption of Beef line:

$$E(y) = 23.5714 + 69.0000(1) + 7.7500x_2 - 12.2857 \times (1)x_2 = 92.5714 - 4.5357x_2.$$

Each one year increase is associated with a decrease of 4.5357 pounds (per person) of beef consumption.

Figure 3.12 shows the regression line for each level.

```
plot(x2,y,  col = "blue", xlab="year",
    ylab="y: meat in pounds (per person)",
    main="Meat Consumption in 7 Years", col.main="darkgreen")
curve(23.5714 +  7.7500*x, from=1, to=10, add=TRUE,
      col="red",lwd=2)
curve(92.5714 - 4.5357*x, from=1, to=10,  add=TRUE,
      col="green",lwd=2)
legend('bottomright',inset=0.05,c("chicken: y=23.5714+7.7500x",
      "beef: y=92.5714-4.5357x"),lty=1,col=c("red","green"))
```

iv)   A point estimate of the average per-person beef consumption in year 8 is given by:

```
newdata=data.frame(x1=1, x2=8, x1x2=8)      #wrap the parameter
predict(Model1,newdata)
```

```
##      1
## 56.29
```

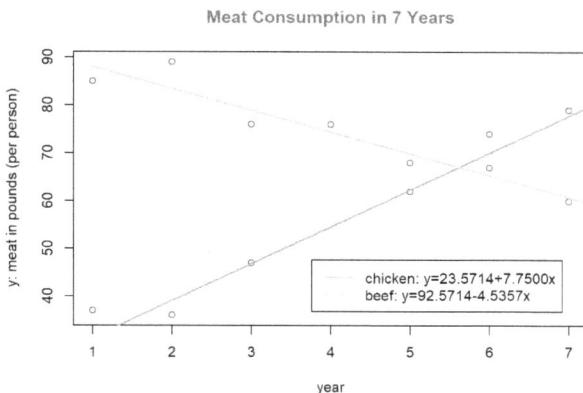

FIGURE 3.12 Linear Model in Meat Consumption

The confidence interval for the mean is given by: $48.43866 < E(y) < 64.13277$.

```
predict(Model1,newdata,level = 0.95,interval="confidence")
```

```
##      fit   lwr   upr
## 1 56.29 48.44 64.13
```

The confidence interval for the predicted value is given by: $44.12911 < y < 68.44232$.

```
predict(Model1,newdata,level = 0.95,interval="predict")
```

```
##      fit   lwr   upr
## 1 56.29 44.13 68.44
```

The two confidence intervals show that it is possible that the consumption of beef declines in year 8.

Sepal Width-Length in Iris data

The iris data, included in the R dataset package, gives the measurements in centimeters of the variables sepal length and width and petal length and width, respectively, for 150 flowers from each of 3 species of iris. The species are Iris setosa, versicolor, and virginica.

  i) Plot the points (Sepal.Width, Sepal.Length) by distinguishing each category using colors.

 ii) Use the ifelse() function in R[1] to define indicator variables and then define the final data frame.

iii) Perform the multilinear regression model describing Sepal.Length mean with respect to Sepal.Width and the categories. Give the equation of the hypothesized line relating the Sepal. Length means the Sepal.Width at each level. Allow interaction between the quantitative and qualitative independent variables.

 iv) Plot these lines on the same graph.

  v) Repeat iii) and iv) for the reduced model that doesn't include the interactions of the variables. Compare the two models.

Solution. i) Figure 3.13 illustrates the desired points.

```
head(iris)
```

```
##    Sepal.Length Sepal.Width Petal.Length Petal.Width
## 1           5.1         3.5          1.4         0.2
## 2           4.9         3.0          1.4         0.2
## 3           4.7         3.2          1.3         0.2
## 4           4.6         3.1          1.5         0.2
## 5           5.0         3.6          1.4         0.2
## 6           5.4         3.9          1.7         0.4
##    Species
## 1   setosa
```

---

[1]statology.org/dummy-variables-in-r/

```
## 2   setosa
## 3   setosa
## 4   setosa
## 5   setosa
## 6   setosa
```

```
plot(iris$Sepal.Width,iris$Sepal.Length, main="Sepal Length versus width",
      pch = 19, col = iris$Species)

legend(x = 3.9, y = 7.8, legend = levels(iris$Species),
      col = 1:length(levels(iris$Species)), pch = 19)
```

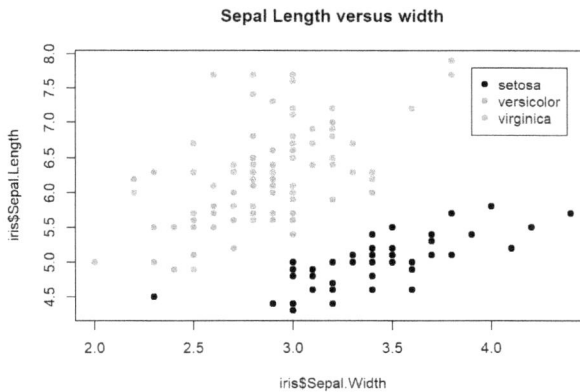

FIGURE 3.13 Iris-Sepal size

ii)

```
# Create indicator variables for setosa and versicolor

Setosa = ifelse(iris$Species == "setosa", 1,  0)
Versicolor= ifelse(iris$Species == "versicolor", 1,  0)

# Create data frame to use for reggression

df_reg = data.frame(Sepal.Length = iris$Sepal.Length,
Sepal.Width = iris$Sepal.Width, Petal.Length = iris$Petal.Length,
Petal.Width = iris$Petal.Width, Setosa=Setosa, Versicolor=Versicolor)

# View the data frame

head(df_reg)
```

```
##    Sepal.Length Sepal.Width Petal.Length Petal.Width
## 1           5.1         3.5          1.4         0.2
## 2           4.9         3.0          1.4         0.2
## 3           4.7         3.2          1.3         0.2
## 4           4.6         3.1          1.5         0.2
## 5           5.0         3.6          1.4         0.2
## 6           5.4         3.9          1.7         0.4
##    Setosa Versicolor
## 1      1          0
## 2      1          0
## 3      1          0
## 4      1          0
## 5      1          0
## 6      1          0
```

iii)

```
# Reading the data

y = iris$Sepal.Length
x = iris$Sepal.Width
x1=Setosa
x2=Versicolor
xx1=x*x1
xx2=x*x2

# Create data frame to use for regression
# that doesn't include Petal Length and Width.

DF=data.frame(y, x, x1, x2, xx1, xx2 )

# create regression model
model = lm(y ~ x + x1 + x2 + xx1 + xx2 )

# View regression model output
summary(model)
```

```
##
## Call:
## lm(formula = y ~ x + x1 + x2 + xx1 + xx2)
##
## Residuals:
##     Min      1Q  Median      3Q     Max
## -1.2607 -0.2586 -0.0331  0.1893  1.4492
##
```

```
## Coefficients:
##                Estimate Std. Error t value Pr(>|t|)
## (Intercept)     3.9068     0.5827    6.71  4.2e-10 ***
## x               0.9015     0.1948    4.63  8.2e-06 ***
## x1             -1.2678     0.8162   -1.55    0.12
## x2             -0.3671     0.8068   -0.46    0.65
## xx1            -0.2110     0.2558   -0.83    0.41
## xx2            -0.0365     0.2793   -0.13    0.90
## ---
## Signif. codes:
## 0 '***' 0.001 '**' 0.01 '*' 0.05 '.' 0.1 ' ' 1
##
## Residual standard error: 0.44 on 144 degrees of freedom
## Multiple R-squared:  0.727,  Adjusted R-squared:  0.718
## F-statistic: 76.9 on 5 and 144 DF,  p-value: <2e-16
```

The fitted regression is:

$$E(y) = 3.90684 + 0.90153x - 1.26784x_1 - 0.36710x_2 - 0.21104xx_1 - 0.03646xx_2$$

Interpretation of the regression coefficients

Only the variable x = Sepal.Width is statistically significant with p-value less than 0.05. The coefficient of determination R^2 shows that 72.74% of the total variation is explained by the regression model. The model fits the data well.

- Substituting $x_1 = 1$ and $x_2 = 0$ into the equation, we obtain the mean Sepal.Length associated with the *Setosa level*:

$$E(y) = 3.90684 + 0.90153x - 1.26784 - 0.21104x = 2.639 + 0.69049x.$$

Each one unit increase in Sepal.Width is associated with an increase of 0.69049 units in Sepal.Length.

- Substituting $x_1 = 0$ and $x_2 = 1$ into the equation, we obtain the mean Sepal.Length associated with *Versicolor level*:

$$E(y) = 3.90684 + 0.90153x - 0.36710 - 0.03646x = 3.53974 + 0.86507x.$$

Each one unit increase in Sepal.Width is associated with an increase of 0.86507 units in Sepal.Length.

• Substituting $x_1 = 0$ and $x_2 = 0$ into the equation, we obtain the mean Sepal.Length associated with *Virginica level*:

$$E(y) = 3.90684 + 0.90153x.$$

Each one unit increase in Sepal.Width is associated with an increase of 0.90153 units in Sepal.Length.

iv)   Figure 3.14 shows the regression line for each level.

```
plot(iris$Sepal.Width,iris$Sepal.Length,
    main= "relationship between Sepal Length and width",
        col.main="darkgreen", pch = 19, col=iris$Species)

legend(x = 2, y = 7.8, legend = levels(iris$Species),
        col = 1:length(levels(iris$Species)), pch = 19)
curve(3.53974 + 0.86507*x, from=1, to=10, add=TRUE,
        col="red",lwd=2)
curve(2.639 + 0.69049*x, from=1, to=10,  add=TRUE,
        col="black",lwd=2)
curve(3.90684 + 0.90153*x, from=1, to=10,  add=TRUE,
        col="green",lwd=2)
```

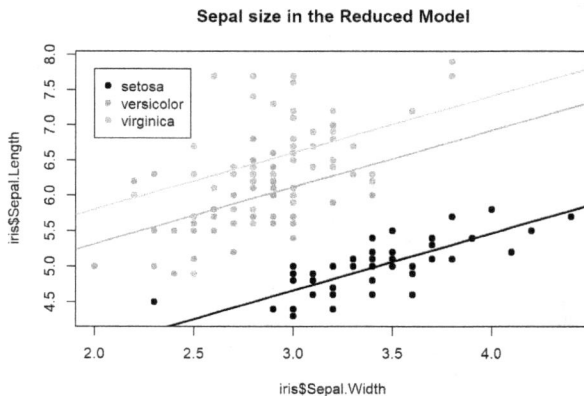

FIGURE 3.14 Linear Model in Sepal size

3.2 Quantitative and Qualitative Predictors in a Regression Model   203

This single model may not be the best fit. However, one can see that Sepal Width-Length relationship behaves differently in the setosa, versicolor, and virginica categories.

v)

▶ **Find the reduced model**

```
Model_R = lm(y ~ x+ x1 + x2, data=DF)
```

▶ **Significance of the coefficients.**

```
summary(Model_R)
```

```
##
## Call:
## lm(formula = y ~ x + x1 + x2, data = DF)
##
## Residuals:
##      Min      1Q  Median      3Q     Max
## -1.3071 -0.2571 -0.0533  0.1954  1.4125
##
## Coefficients:
##               Estimate Std. Error t value Pr(>|t|)
## (Intercept)    4.1982     0.3223   13.03  < 2e-16 ***
## x              0.8036     0.1063    7.56  4.2e-12 ***
## x1            -1.9468     0.1000  -19.47  < 2e-16 ***
## x2            -0.4881     0.0902   -5.41  2.5e-07 ***
## ---
## Signif. codes:
## 0 '***' 0.001 '**' 0.01 '*' 0.05 '.' 0.1 ' ' 1
##
## Residual standard error: 0.438 on 146 degrees of freedom
## Multiple R-squared:  0.726,  Adjusted R-squared:  0.72
## F-statistic:  129 on 3 and 146 DF,  p-value: <2e-16
```

The p-values is less than 0.05 for each variable. Therefore, each variable is statistically significant.

▶ **ANOVA** *F*-test for the Reduced Model

```
Model0 = lm(y ~ 1, DF)
anova(Model0, Model_R)
```

```
## Analysis of Variance Table
##
## Model 1: y ~ 1
## Model 2: y ~ x + x1 + x2
##   Res.Df RSS Df Sum of Sq     F Pr(>F)
## 1    149 102
## 2    146  28  3       74.2 129 <2e-16 ***
## ---
## Signif. codes:
## 0 '***' 0.001 '**' 0.01 '*' 0.05 '.' 0.1 ' ' 1
```

From the *p*-value $< 2.2 \times 10^{-16}$ of the *F*-test, the reduced model fits the data well.

▶ **Comparing the Reduced and the Complete Models**

We perform an *F*-test for the following null Hypothesis H_0 versus the alternative Hypothesis H_a:

$$H_0: \quad \beta_4 = \beta_5 = 0 \qquad \text{versus}$$

$$H_a: \quad \text{at least one of the parameters } \beta_4, \beta_5 \text{ differs from 0.}$$

```
Model_C = model            # the complete model
anova(Model_R, Model_C)
```

```
## Analysis of Variance Table
##
## Model 1: y ~ x + x1 + x2
## Model 2: y ~ x + x1 + x2 + xx1 + xx2
##   Res.Df  RSS Df Sum of Sq     F Pr(>F)
## 1    146 28.0
## 2    144 27.9  2     0.157 0.41   0.67
```

From the "anova(Model_R, Model_C)", we read a large p-value $= 0.6668$. Therefore, we cannot reject H_0.

▶ **Plotting the data and the lines at each level of the reduced model**

We have the following equations of the mean Sepal.Length associated with:

- *Reduced model*: $E(y) = 4.19821 + 0.80356x - 1.94682x_1 - 0.48807x_2$.
- *Setosa level* $(x_1 = 1, x_2 = 0)$: $E(y) = 2.25139 + 0.80356x$.
- *Versicolor level* $(x_1 = 0, x_2 = 1)$: $E(y) = 3.71014 + 0.80356x$.
- *Virginica level* $(x_1 = 0, x_2 = 0)$: $E(y) = 4.19821 + 0.80356x$.

Note the three lines have the same slope. Thus they are parallel (see Figure 3.15).

```
plot(iris$Sepal.Width,iris$Sepal.Length,
     main= "Sepal size in the Reduced Model",
     col.main="darkgreen", pch = 19, col=iris$Species)

legend(x = 2, y = 7.8, legend = levels(iris$Species),
       col = 1:length(levels(iris$Species)), pch = 19)
curve(3.71014 + 0.80356*x, from=1, to=10, add=TRUE,
      col="red",lwd=2)
```

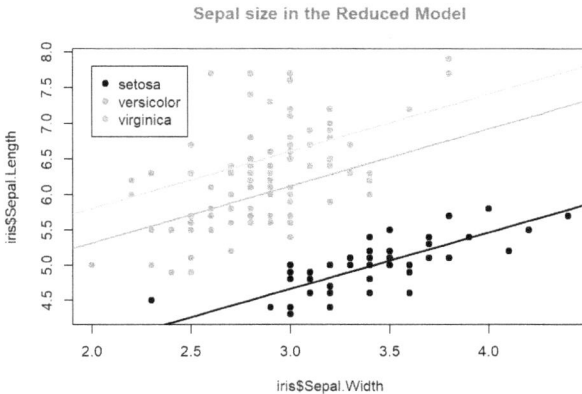

FIGURE 3.15 Reduced Linear Model in Sepal size

```
curve(2.25139 + 0.80356*x, from=1, to=10,  add=TRUE,
      col="black",lwd=2)
curve(4.19821 + 0.80356*x, from=1, to=10,  add=TRUE,
      col="green",lwd=2)
```

Appendix: Q-Q plot

There exist several methods[2] for testing the normality distribution for a given data set. An assumption of normality was made for the residuals in the multi-linear regression model. This assumption needs to be satisfied to evaluate the efficiency of the model.

We describe here a visual method by creating a Q-Q plot. Our conclusion about normality with this plot will be subjective as a result of a visual inspection.

Q-Q plot : Description

The Q-Q plot[3], where Q stands for quantile, is a graphical technique for determining if two data sets come from populations with a common distribution.

A Q-Q plot is a plot of the quantiles of the first data set against the quantiles of the second data set.

A **quantile** is a fraction or percentage of points below a given value.

---

[2]https://www.statology.org/test-for-normality-in-r/
[3]https://www.itl.nist.gov/div898/handbook/eda/section3/qqplot.htm

In particular, if we want to determine that a data set comes from a population normally distributed, we check whether

$$P_{unknown}(X \leqslant x) = P_{unknown\ standardized}(Z = \frac{X - \mu}{s} \leqslant \frac{x - \mu}{s})$$

$$\approx P_{normal}(Z = \frac{X - \mu}{s} \leqslant z)$$

For this purpose, assume we can find an approximation of the unknown probability:

$$P_{unknown}(X \leqslant x) = P_{unknown}(Z = \frac{X - \mu}{s} \leqslant \frac{x - \mu}{s}) \approx p_x$$

then, if the *unknown* distribution is approximately *normal*, we would have:

$$P_{unknown}(Z = \frac{X - \mu}{s} \leqslant \frac{x - \mu}{s}) \approx p_x \approx P_{normal}(Z = \frac{X - \mu}{s} \leqslant z).$$

Each quantile z is obtained by solving the equation

$$P_{normal}(Z \leqslant z) = p_x \approx P_{unknown}(Z \leqslant \frac{x - \mu}{s})$$

from which we deduce that

$$\frac{x - \mu}{s} \approx z.$$

A plot of the quantiles-quantiles (z, x) would show aligned points.

Deviations from a straight line suggest departures from normality.

Note that general methods to numerically compute the quantiles for general classes of distributions are built into many statistical software packages.

Cumulative distribution function (cdf): F

When we don't know the distribution of the sample, we will use an approximation of the unknown probability.

First, we show how a quantile relates to a probability (R. Bartoszynski, 2008).

Definition. For any random variable (r.v) X, the function of real variable t defined by

$$F(t) = P(X \leqslant t)$$

is called the cumulative probability distribution function (cdf) of X.

Theorem. Every cdf F has the following properties:

i) F is nondecreasing.

ii) $\lim\limits_{t \to -\infty} F(t) = 0,$ $\lim\limits_{t \to +\infty} F(t) = 1$

iii) F is right continuous.

Definition. Let X be a r.v with a cdf F. For $p \in (0,1)$, we define the pth quantile of X as any solution of the inequations

$$P(X \leqslant x) \geqslant p, \qquad\qquad P(X \geqslant x) \geqslant 1 - p$$

or equivalently

$$F(x) \geqslant p, \qquad\qquad F(x_-) \leqslant p$$

If F is continuous, then for a p percentile, the pth quantile x is the solution of the equation
$$F(x) = P(X \leqslant x) = p.$$

Next, we give the general theory that justifies the approximation of the unknown probability (R. Bartoszynski, 2008).

Empirical cumulative distribution function (ecdf): F_n

Let $X_1, \ldots, X_n$ be independent and identically distributed random variables (iid); that is $X_1, \ldots, X_n$ is a random sample.

Let $F(t) = P(X_i \leqslant t)$ be the common cdf of the X_i.

For n and t, the empirical cumulative distribution function (ecdf) of the sample $X_1, \ldots, X_n$ is defined by

$$F_n(t) = \frac{\text{number of data values that do not exceed } t}{n}$$

The ecdf $F_n(t)$ is formed by ordering the set of sample values from the smallest to largest.

Properties of F_n:

 — $F_n(t)$ is a step function that increases by a multiple of $1/n$ at each point of the sample.

 — $F_n(t)$ is a random variable that depends on the n observations $X_1, \ldots, X_n$.

 — $nF_n(t)$ is a random variable, equal to the number of X_i's among the n observations $X_1, \ldots, X_n$. Therefore, $nF_n(t)$ is $\mathcal{B}(n, p)$ with $p = P(X_i \leqslant t) = F(t)$.

 — By the Law of Large Numbers, we have

$$F_n(t) = \frac{nF_n(t)}{n} \quad \longrightarrow \quad F(t) \qquad \text{with probability 1.}$$

 — By Glivenko-Cantelli Theorem, we have

$$\sup_t |F_n(t) - F(t)| \quad \longrightarrow \quad 0 \qquad \text{with probability 1.}$$

The convergences above justify the use of the ecdf F_n as an approximation of the unknown edf F.

How to create a Q-Q plot ?

▶ Empirical cumulative distribution function[4] F_n (ecdf) and Q-Q plots are used to visualize data by highlighting properties of the distribution rather than the individual points.

ecdfs and Q-Q plots are based on ranking the data and visualizing the relationship between ranks and the actual values.

The graph of the ecdf plots the points $(x, F_n(x))$.

The Q-Q plot plots the points (Zar, 1984)

$$\Big(Theoretical\ quantile = z = qnorm(F_n(x)), \qquad sample\ quantile = x\Big)$$

where *qnorm* is the R function that calculates the quantile corresponding to the probability $F_n(x)$. The points (z, x) will fall approximately on a line if the data comes from an approximately normally distributed population.

▶ Different sources use slightly different approximations of the cumulative distribution.

The formula[5] used by the "qqnorm" function in the basic "stats" package in R (programming language) is as follows:

$$z_i = \Phi^{-1}\Big(\frac{i-a}{n+1-2a}\Big) \quad i = 1, 2, \ldots, n, \qquad a = \begin{cases} 3/8, & \text{if } n \leqslant 10 \\ \\ 1/2, & \text{if } n > 10, \end{cases}$$

where i serves to index the position of the data points x_i in the ordered sample data set, and Φ^{-1} is the standard normal quantile function[6].

---

[4]https://clauswilke.com/dataviz/ecdf-qq.html
[5]https://statproofbook.github.io/P/norm-qf.html
[6]https://en.wikipedia.org/wiki/Normal_probability_plot

Normality check in the mtcars model

▶   Normality check in the mtcars model using F_n

```
# Select a subset from mtcars data set
DF = subset(mtcars , select=-c(cyl, vs, am, gear,carb))

# Model4 for mtcars data
Model4 = lm(mpg ~  wt + qsec , data=DF)

# The observed residuals
residuals4 = DF$mpg - ( coefficients(Model4)[1]
            + coefficients(Model4)[2]*mtcars$wt
            + coefficients(Model4)[3]*mtcars$qsec )
```

```
m=mean(residuals4)           # The mean of the residuals in Model4
m
```

```
## [1] 7.605e-15
```

Note that the mean of the residuals is equal to 0.

```
s=sd(residuals4)             # standard deviation of the residuals
xstand = (residuals4-m)/s    # standarized residuals
x=xstand
```

```
Fn=ecdf(x)
# Fn is the ecdf function defined using the set of data xstand
# Fn estimates the unknown probabilities
# Fn(xtand) returns the percentiles for xstand
```

```
# z-quantiles for the calculated percentiles Fn(xstand)
z=qnorm(Fn(xstand), mean=0, sd=1)

par(mfrow=c(1,2))
plot(xstand, Fn(xstand), main="plot of the ecdf Fn")
plot(z, xstand, main="observed vs theoretical quantiles")
```

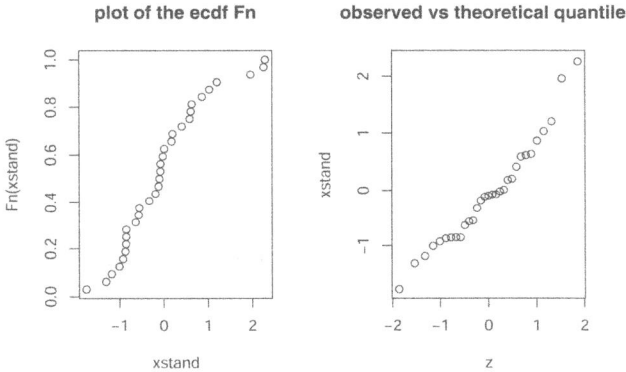

FIGURE 3.16 Residuals: ecdf and q-q plot

Note that the graph shows an approximate linear relationship:

$$z \approx (residuals4 - m)/s.$$

We used a ready ecdf function implemented in R. One can define such a function using a for loop[7]

```
# Define a function to compute the ECDF
ecdf_func <- function(data) {
    Length <- length(data)
    sorted <- sort(data)

    ECDF <- rep(0, Length)
    for (i in 1:Length) {
        ECDF[i] <- sum(sorted <= data[i]) / Length
    }
```

---

[7]https://www.geeksforgeeks.org/compute-empirical-cumulative-distribution-function-in-r/

```
    return(ECDF)
}
```

```
ecdf1 = ecdf_func(xstand)
```

```
# Check whether ecdf_func plays the same role as ecdf.
identical(ecdf1, Fn(xstand))
```

```
## [1] TRUE
```

▶   Normality check in the mtcars model using qqnorm

```
par(mfrow=c(1,2))

#Plot the Histogram of the residuals
hist(xstand, col="lightblue",main="Distribution of the residuals",
     xlab ="residuals", freq=FALSE,breaks=8)

curve(dnorm( x, mean=0, sd=1 ), add=TRUE, col="red", lwd=2)

#Plot the Q-Q plot of the standardized residuals
qqnorm(xstand)
qqline(xstand, col="red", lwd=2)
```

Using F_n and qqnorm, we observe that most of the points fall close to a line. The normality assumption is reasonable. Note also that the two graphs of quantiles-quantiles are approximately similar (Figure 3.17).

▶   Remark

The visual inspection performed through the Q-Q plot is usually unreliable. But statistical programs on computers make it possible with a little effort. It is an important first step and should not be omitted. If the graph reveals any deviations from normality or if any other assumption for regression is not satisfied, a different model may be considered.

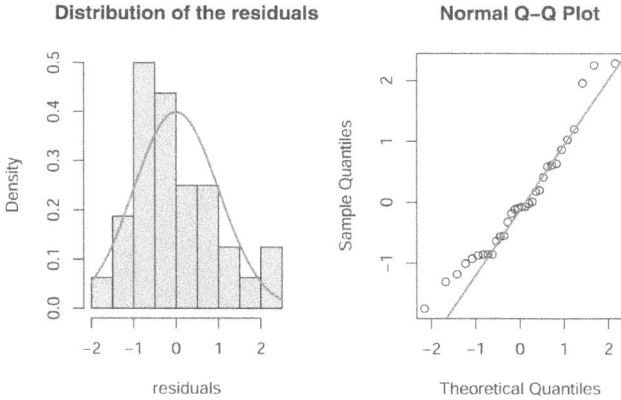

Distribution of the residuals Normal Q–Q Plot

FIGURE 3.17 Residuals: check normality

When the visual inspection doesn't reveal a departure from assumptions, it is possible to use a significance test in order to ascertain whether the data shows or not a serious deviation from normality or not. For example, the Kolmogorov-Smirnov test considers the null hypothesis:

$$H_0 : \quad F = F_0 \qquad \text{versus the alternative hypothesis} \qquad H_a : \quad F \neq F_0,$$

where the hypothetical cdf F_0 is from a normal one. The statistic involved, if H_0 is true, is

$$D = \max_t |F_n(t) - F_0(t)| \qquad \text{with}$$

$$\frac{F_n(t) - F_0(t)}{\sqrt{\frac{F_n(t)(1 - F_n(t))}{n}}} \quad \leadsto \quad \mathcal{N}\left(0, 1\right) \quad \text{for } n \text{ large.}$$

This test is performed in R, for the Model 4-mtcars data set, as follows:

```
ks.test(xstand, z )


##
##   Two-sample Kolmogorov-Smirnov test
##
## data:  xstand and z
## D = 0.12, p-value = 1
## alternative hypothesis: two-sided
```

From the output, the p-value is greater than 0.05, implying that the distribution of the data is not significantly different from normal distribution. In other words, we can assume the normality.

Appendix: Residual Plots

Examining the errors or residuals

$$\epsilon_i = \text{observed } y_i - \text{ predicted } y_i = y_i - \widehat{y}_i$$

helps to assess how well the approximate model describes the data since they show how far the data fall from the regression model. In the multilinear relationship, the mean of the least squares residuals is zero $E(\epsilon_i) = 0$. Thus the residual plot; that is, the scatterplot of the regression residuals against the explanatory variables, informs us how the points locate about the horizontal line $residual = \epsilon = 0$.

On the other hand, if the regression model catches the overall pattern of the data, there should be no pattern in the residuals (David S. Moore, 2003). It will be an unstructured horizontal band centered at zero and symmetric about zero.

It is usual to plot the residuals versus the predicted values $\widehat{y}$ and also versus each of the explanatory variables.

Residual Plots in the mtcars model

```r
# Select a subset from mtcars data set
DF = subset(mtcars , select=-c(cyl, vs, am, gear,carb))

# Model4 for mtcars data
Model4 = lm(mpg ~  wt + qsec , data=DF)

# The predicted values using Model4
yhat = ( coefficients(Model4)[1]
           + coefficients(Model4)[2]*mtcars$wt
           + coefficients(Model4)[3]*mtcars$qsec )
# The observed residuals
residuals4 = DF$mpg - yhat
```

```r
# regrouping 3 plots in one row

par(mfrow=c(1,3))

# Plot the horizontal line of residuals = 0
# and the residuals versus "wt"

plot(mtcars$wt,residuals4, main="residuals versus wt")
curve(0*x, from=-30, to =30, add=TRUE, col="red", lwd=3)

# Plot the line residuals = 0
# and the residuals versus the variable "qsec"

plot(mtcars$qsec,residuals4, main="residuals versus qsec")
curve(0*x, add=TRUE, col="red", lwd=3)

# Plot the line residuals = 0
# and the residuals versus the predicted values yhat

plot(yhat,residuals4, main="residuals versus yhat")
curve(0*x, add=TRUE, col="red", lwd=3)
```

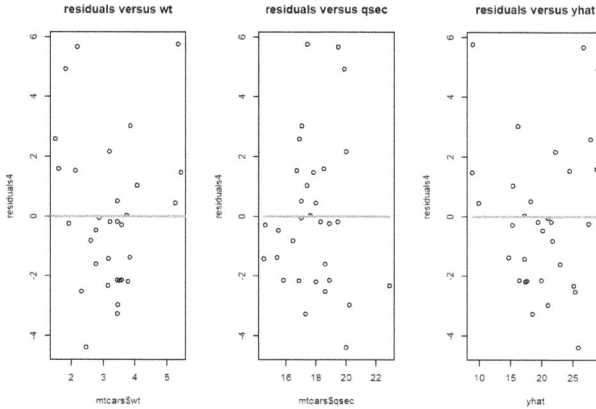

FIGURE 3.18 Residuals versus explanatory variables

```
mean(residuals4)
```

```
## [1] 7.605e-15
```

```
sd(residuals4)
```

```
## [1] 2.511
```

The residuals in Figure 3.18 appear to be randomly scattered and exhibit no obvious anomalies or patterns. They also satisfy the assumption of mean zero for this type of model. They are located within three standard deviations, $3s \approx 7.5$. Thus, if the assumption of normality is satisfied, one can say that there is no outlier.

Bibliography

A. Gelman (2006). *Data Analysis Using Regression and Multilevel Hierarchical Models*. Analytical Methods for Social Research. Cambridge University Press.

Challal, S. (2020). *Introduction to the theory of optimization in Euclidean space*. CRC Press. Taylor & Francis Group.

Chang, W. (2013). *R Graphics Cookbook. Practical Recepies for Visualizing Data*. O'Reilly.

Crawley, M. J. (2013). *The R Book*. A John Wiley & Sons, Ltd., Publication, 2nd edition.

David S. Moore (2003). *Introduction to the practice of statistics*. W. H. Freeman and Company, 4 edition.

D.D. Wackerly, W. Mendenhall (2008). *Mathematical Statistics with Applications*. Brooks/Cole Cengage Learning, 7th edition.

Gardener, M. (2012). *Beginning R. The Statistical Programming Language*. John Wiley & Sons, Inc.

Gut, A. (1995). *An intermediate course in probability*. Spring-Verlag.

Holt, J. (2013). *Linear Algebra with Applications*. W. H. Freeman and Company.

James T. McClave (2013). *Statistics*. Pearson, 12 edition.

K. Sydsæter, P. Harmmond (2004). *Further Mathematics for Economic Analysis*. Wiley Series in Probability and Statistics, 2nd edition.

Lander, J. P. (2014). *R for Everyone. Advanced Analytics and Graphics*. Addison Wesley Data & Analytics Series.

Larson, H. J. (1982). *Introduction to probability theory and statistical inference*. John Wiley & Sons.

P. Roback (2021). *Beyond Multiple Linear Regression. Applied Generalized Linear Models and Multilevel Models in R*. CRC Press. Taylor & Francis Group.

Panik, M. J. (2010). *Regression Modeling. Methods, Theory, and Computation with SAS*. CRC Press. Taylor & Francis Group.

Panik, M. J. (2012). *Statistical Inference. A Short Course.* CRC Press. Wiley.

posit.co/download/rstudio desktop (Accessed on 2025-02-24). *RStudio Desktop.*

R. Bartoszynski (2008). *Probability and Statistical Inference.* Wiley-Interscience, 2nd edition.

r project.org (Accessed on 2025-02-24). *The RProject for Statistical Computing.*

R.J. Larsen (2001). *An introduction to mathematical statistics and its applications.* Prentice Hall, 3rd edition.

rmarkdown.rstudio.com (Accessed on 2025-02-24). *R Markdown from R Studio.*

S. Chatterjee (2013). *Handbook of Regression Analysis.* A John Wiley & Sons, Inc., Publication.

S. Dowdy, S. Wearden (2004). *Statistics for Research.* Wiley Series in Probability and Statistics, 3rd edition.

STHADA (Accessed on 2025-02-24). *R Basics: Quick and Easy.* Statistical tools for high-throughput analysis.

William Mendenhall, Robert J. Beaver (2018). *Introduction to probability and statistics.* Cengage and Learning, 15 edition.

Xie, Y. (2015). *Dynamic Documents with R and knitr.* Chapman and Hall/CRC, Boca Raton, Florida, 2nd edition. ISBN 978-1498716963.

Xie, Y. (2024). *bookdown: Authoring Books and Technical Documents with R Markdown.* CRC Press. A Chapman & Hall Book.

Yihui Xie, J. J. Allaire (2023). *R Markdown: The Definitive Guide.* CRC Press. A Chapman & Hall Book.

Zar, J. H. (1984). *Biostatistical Analysis.* Prentice Hall.

Index